# Strategic Writing

This practical, multidisciplinary text teaches high-quality public relations and media writing with clear, concise instructions for more than 40 types of documents.

*Strategic Writing* takes a reader-friendly "recipe" approach to writing in public relations, advertising, sales and marketing, and other business communication contexts, illustrated with examples of each type of document. With concise chapters on topics such as ethical and legal aspects of strategic writing, including diversity and inclusion, this thoroughly updated fifth edition also includes additional document samples and coverage of writing for various social media platforms. Packed with pedagogical resources, *Strategic Writing* offers instructors a complete, ready-to-use course.

It is an essential and adaptable textbook for undergraduate courses in public relations, advertising and strategic communication writing, particularly those that take a multidisciplinary and multimedia approach.

*Strategic Writing* is ideally suited for online courses. In addition to syllabi for both online and traditional courses, the instructor's manual includes Tips for Teaching Strategic Writing Online. Those tips include easy guidelines for converting the book's PowerPoint slides to videos with voiceovers for online lectures. The book's recipe-with-examples approach enhances student self-instruction, particularly when combined with the companion website's sample assignments and grading rubrics for every document. Visit the site at www.routledge.com/cw/marsh.

**Charles Marsh** is the Oscar Stauffer Professor of Journalism and Mass Communications at the University of Kansas. His primary areas of research involve public relations, theories of cooperation and classical rhetoric.

**David W. Guth** is Associate Professor Emeritus at the William Allen White School of Journalism and Mass Communications, University of Kansas. His areas of special research interest are crisis communication, political communication and public relations history.

**Bonnie Poovey Short** was founder of Short Solutions, an award-winning editorial and creative services firm that specialized in the healthcare field. Though now retired, she also taught at the university level and served as communications coordinator for a school district.

"This text is a comprehensive guide of persuasive writing techniques for advertising, public relations, marketing, and business writers. It's an ideal learning tool for beginners as well as a reference for advanced writers. The book is unique in its approach of presenting small, digestible chunks of different types of writing styles illustrated with lots of examples. It includes thoughtful sections on writing for diverse audiences. The updated sections on writing for social media feature practical tips beyond writing, including when and how often to post. *Strategic Writing* should be on every writer's bookshelf!"

– Frauke Hachtmann, University of Nebraska—Lincoln

# Strategic Writing

## Multimedia Writing for Public Relations, Advertising and More

**Fifth Edition**

Charles Marsh
David W. Guth
Bonnie Poovey Short

NEW YORK AND LONDON

Fifth edition published 2021
by Routledge
52 Vanderbilt Avenue, New York, NY 10017

and by Routledge
2 Park Square, Milton Park, Abingdon, Oxon, OX14 4RN

*Routledge is an imprint of the Taylor & Francis Group, an informa business*



First edition published by Pearson Education Inc. 2005
Fourth edition published by Routledge 2018

*Library of Congress Cataloging-in-Publication Data*
Names: Marsh, Charles, 1955– author. | Guth, David, author. | Short, Bonnie Poovey, author.
Title: Strategic writing : multimedia writing for public relations, advertising and more / Charles Marsh, David W. Guth, Bonnie Poovey Short.
Description: 5. | New York, NY : Routledge, 2021. | Includes bibliographical references and index.
Identifiers: LCCN 2020027194 (print) | LCCN 2020027195 (ebook) | ISBN 9780367895396 (hardback) | ISBN 9780367895402 (paperback) | ISBN 9781003019701 (ebook)
Subjects: LCSH: Business writing—Problems, exercises, etc. | Business report writing—Problems, exercises, etc.
Classification: LCC PE1479.B87 M35 2021 (print) | LCC PE1479.B87 (ebook) | DDC 808.06/665—dc23
LC record available at https://lccn.loc.gov/2020027194
LC ebook record available at https://lccn.loc.gov/2020027195

ISBN: 978-0-367-89539-6 (hbk)
ISBN: 978-0-367-89540-2 (pbk)
ISBN: 978-1-003-01970-1 (ebk)

Typeset in Optima
by Apex CoVantage, LLC

Access the companion website: www.routledge.com/cw/marsh

# Contents

# Introduction How This Book Works

## New to This Edition

New technology continues to revolutionize the world of strategic communication. The fifth edition of *Strategic Writing: Multimedia Writing for Public Relations, Advertising and More* reflects these changes. In addition to a line-by-line revision and updated examples, this edition features three new sections:

- Social media calendars
- Creative briefs
- Promotional news releases

Based on new research and requests from professors who use this textbook, we've also updated and often expanded instructions for every document in this book. Digital communications and social media are transforming the media of strategic writing, and this new edition reflects that reality. (Actually, we had planned to bullet-point *all* the updates and expansions here, but the list is just too long. Every document from news releases to strategic message planners to marketing communication plans has been updated to reflect current usage.)

One thing that *hasn't* changed is the book's closing focus on grammar and style (although we have revised the book's grammar and style appendices). It remains our hope that the appendices will allow a professor simply to type "PM3," for example, on a student's assignment rather than explaining at length the wondrous nature of the comma splice.

Once upon a time (longer ago than we want to admit), the three authors of this book were young professionals. Our careers took us through jobs with corporations, government bureaus, nonprofit organizations and marketing agencies in addition to freelance work. Then we went crazy (according to some) and decided to become professors. As professors, we quickly learned three things about textbooks that teach writing for such careers. Although there are many good books,

- Most tend to cover only one area of writing, such as public relations or advertising.
- They're long.
- Being long, they don't always give concise instructions about how to write particular documents.

As professionals, we hadn't been limited to just one area of writing. All of us wrote speeches (public relations), advertisements (advertising), brochures (sales and marketing), business reports (business communication) and more. We wrote for print, for video and, more recently, for online and social media. We believe that kind of multidisciplinary, multimedia convergence is standard for today's writers. We wanted a book that prepared students for that diversity.

As professionals, we worked on tight deadlines. Therefore, we wanted a book that got right to the point. For example, if your professor assigns a social media calendar that's due tomorrow, we wanted you to have a book that offered a clear set of instructions—a recipe, basically—for writing that document. We wanted a book that you could use as a desktop reference—either digital or paper—while you wrote. And we wanted a book that you could take to the job after graduation.

Finally, we wanted a book that emphasized the importance of strategy. As young professionals, we weren't always sure *why* we were writing *what* we were writing. Sometimes we wrote documents just because the boss told us to; we didn't always examine *why* we were writing those documents. We didn't always wonder *how* those documents helped move our organization toward the achievement of its goals. We want you to be better than that. We want you to realize that, ideally, every business document you write—whether it's a business report or a tweet—advances the organization toward a specific business goal. We want you to understand that every document should aim at a desired effect. We even want to help you identify what the goal-oriented message of each document should be. Our name for this goal-oriented writing is *strategic writing*. We describe strategic writing in Section 1.

## How This Book Is Organized

We've divided this book into five sections:

- The first section gives you the background on strategic writing. It's packed with information, but we've kept each segment short.
- The second section gives instructions for public relations documents.
- The third section gives instructions for advertising documents.
- The fourth section gives instructions for sales and marketing documents.
- The fifth section gives instructions for business communication documents.

Of course, some documents—podcasts, brochures and tweets, for example—could fit into each of these four professional categories. The book concludes with four appendices that offer guidelines for punctuation, grammar, style and oral presentations.

Another useful way to categorize strategic communication documents is to understand the four kinds of media in which they appear. You can use the PESO acronym to remember paid, earned, shared and owned. These are the four traditional paths to reaching target audiences:

- Paid media generally are outlets, such as YouTube or a paper/digital newspaper, that accept advertising.
- Earned media generally are news outlets that accept news releases and similar offerings—if those offerings are approved by the medium's gatekeeping editors.
- Shared media generally are social media, in which we share information and opinions with others.
- Owned media generally are those controlled by an organization, such as a company's digital newsroom.

Just like advertising, public relations, sales and other areas of strategic communication, these four kinds of media can overlap with one another. YouTube, for example, can be paid or shared, depending on what type of message we're using it to send.

We hope the purpose of this book is clear: It explains strategic writing. It offers instructions—recipes—for dozens of different strategic documents. And it offers help with punctuation, grammar and style. It's designed to be a user-friendly tool that helps make you an effective writer.

## Acknowledgments

The authors wish to thank the many people who helped make this book possible, including Dean Ann Brill and the faculty, staff and students of the William Allen White School of Journalism and Mass Communications at the University of Kansas.

As the book progressed from draft to draft, we greatly appreciated the educators and professionals who generously offered advice: Sarah Crowther, Swansea University; Frauke Hachtmann, University of Nebraska-Lincoln; Jordan Stalker, DePaul University; Hilary Stepien, Bournemouth University; and Jennifer Wood, Millersville University. In moving from manuscript to the book you're now reading, we relied on the indispensable expertise of the team at Routledge and Apex CoVantage, particularly Brian Eschrich, Grant Schatzman, Kate Fornadel and copy editor extraordinaire Ginjer Clarke.

And, as always, the authors thank their families for their love and patience.

Part of good communication is feedback, and we invite you to write us with questions or comments on this book.

Charles Marsh
University of Kansas
marsh@ku.edu

David W. Guth
University of Kansas
dguth@ku.edu

Bonnie Poovey Short
Short Solutions
Chapel Hill, NC
bpshort105@gmail.com

# Appendices Summaries

The full appendices begin on page 333.

## *Punctuation Rules: Appendix A*

**PM1** Commas with coordinating conjunctions
**PM2** No commas with coordinating conjunctions
**PM3** Comma splices
**PM4** Commas with opening subordinate clauses
**PM5** No commas with closing subordinate clauses
**PM6** No commas with restrictive relative clauses
**PM7** Commas with nonrestrictive relative clauses
**PM8** Commas with nonrestrictive nouns
**PM9** Commas with titles
**PM10** No commas with restrictive nouns
**PM11** Commas with dates
**PM12** Commas with state and nation names
**PM13** Commas with adjective series
**PM14** Commas with *and* series
**PM15** Commas with opening quotation marks
**PM16** Commas with attributions after quotations
**PM17** Commas following other punctuation marks
**PM18** Commas with attributions that split quotations
**PM19** No commas with attributions before paraphrased quotations
**PM20** Commas with closing quotation marks
**PM21** Commas with direct address
**PM22** Commas with *yes* and *no*
**PM23** Commas with people's ages
**PM24** Commas with present participial phrases
**PM25** Commas with past participial phrases
**PM26** Commas with opening phrases
**PM27** Commas with opening *-ly* adverbs
**PM28** Commas with *not*

**PM29** Commas with interjections
**PM30** Commas with conjunctive adverbs (*however*, etc.)
**PM31** No commas after opening coordinating conjunctions
**PM32** Commas that incorrectly separate subject and verb
**PM33** Semicolons with independent clauses
**PM34** Semicolons with series
**PM35** Colons only after independent clauses
**PM36** Colons with lists
**PM37** Quotation marks within quotation marks
**PM38** Quotation marks with multi-paragraph quotations
**PM39** Quotation marks in headlines
**PM40** Quotation marks with titles
**PM41** Quotation marks with unfamiliar words
**PM42** Question marks with closing quotation marks
**PM43** Question marks in two-question sentences
**PM44** Question marks that don't end sentences
**PM45** No question marks with statements
**PM46** Exclamation points with closing quotation marks
**PM47** Exclamation points that don't end sentences
**PM48** Periods with closing quotation marks
**PM49** Periods with sentence-ending abbreviations
**PM50** Periods with parentheses
**PM51** Dashes with sentence interruptions
**PM52** Dashes with dramatic pauses
**PM53** Dashes with explanations of words or concepts
**PM54** Dashes as informal colons
**PM55** Hyphens with compound modifiers
**PM56** Hyphens with prefixes
**PM57** Suspensive hyphenation
**PM58** Apostrophes with contractions
**PM59** Apostrophes with possession
**PM60** No apostrophes with personal pronouns
**PM61** No apostrophes with adjectival plural nouns

**PM62** Apostrophes with decade abbreviations

**PM63** Parentheses with nonessential comments

## *Grammar Rules: Appendix B*

**G1** Pronoun disagreement

**G2** False series/lack of parallelism

**G3** Subject–verb disagreement

**G4** *Who* and *whom*

**G5** Dangling participles, dangling modifiers

**G6** *That* and *which*

**G7** *I* and *me*

## *Style Guidelines: Appendix C*

**S1** Business titles

**S2** Abbreviations in company names

**S3** Second reference

**S4** Numbers

**S5** Abbreviation of month names

**S6** State names

**S7** Percentages

**S8** Commas in series

SECTION

# 1

# Strategic Writing

## OBJECTIVES

**In Section 1: Strategic Writing, you will learn what strategic writing is. In doing so, you will learn about:**

- The importance of good writing
- Research, planning and the writing process
- Writing for the ears
- Writing for the web
- Writing for social media
- Strategic design
- Integrated marketing communications
- Ethics and strategic writing
- Diversity and strategic writing
- Persuasion and strategic writing
- The law and strategic writing
- Jobs in strategic writing

# 1A Introduction to Strategic Writing

Strategic writing is goal-oriented writing. Well-managed organizations have specific, written goals they strive to achieve. Strategic writing helps those organizations achieve those goals.

Another title for this book could be *Writing That Helps Organizations Achieve Their Specific Business Goals*—but that's too long. We prefer *Strategic Writing*. In the paragraphs that follow, we'll further explain what we mean by strategic writing. Right now, however, you'll establish a solid foundation in this exciting profession if you focus on the following key idea: Whenever you write a specific document for public relations, advertising, business communication or sales and marketing, you should know the purpose of the document and how that purpose relates to a specific organizational goal. Imagine what a successful strategic writer you'll become if every document you produce moves your organization closer to the fulfillment of its specific goals.

The word *strategy* comes to English from Greek. In Greek, *strategos* means *military leader* or *general*. As a strategic writer, you're like a general: With each document, you direct your ideas, words and other multimedia elements on a specific mission. Most dictionaries define the word *strategy* as meaning something like "planned actions in support of a particular policy." In strategic writing, your "planned actions" involve the process of writing, and the "particular policy" is the business goal you're helping to achieve.

As a strategic writer, you should be familiar with business goals, so let's examine where they come from. Many organizations have a values statement, a brief set of core values that ideally guide their actions. Those general values often lead to a mission statement that more precisely describes the organization's purpose. In order to fulfill that mission, organizations usually create annual business goals.

As a strategic writer, your job is to help your organization achieve its values-driven, mission-related goals. That sounds like a big job—and it is. If you're writing a multimedia newsletter story, you need to know what business goal or goals your story and the newsletter are helping to achieve. The same is true for a mobile ad and a content-marketing video. Ideally, everything you write on behalf of your organization moves it toward the fulfillment of its mission.

Your communication skills will be particularly important to your organization: To achieve their goals, organizations rely partly on resources controlled by other groups.

For example, nonprofit organizations rely in part on the resource of contributions made by donors. For-profit companies rely on the resource of purchases made by customers. Both nonprofit and for-profit organizations depend on the resource of fair coverage that the news media hold. In every organization, top management relies on the resource of productivity held by other employees. Your mission as a strategic writer is to secure those resources through effective communication. Much of your work as a strategic writer will involve building productive relationships with resource holders.

In building those relationships, the documents that you produce can be divided into two groups: on-strategy and off-strategy. On-strategy documents remain focused on a clear message that helps an organization achieve a particular goal or goals. Off-strategy documents fail to connect with business goals. Either they began with no consideration of business goals, or they attempted but failed to create a goal-oriented message. Off-strategy documents can be worse than a waste of time. Their nonstrategic messages can confuse target audiences, who become unsure about what message you're trying to send and what kind of relationship you seek.

There are many ways to categorize strategic messages. In this book, we divide them among the overlapping disciplines of public relations, advertising, business communication and sales and marketing. A different way to categorize them is to identify the four kinds of media in which they appear. You can use the PESO acronym to remember paid, earned, shared and owned media. Those are the four traditional paths to reaching target audiences. For more information on PESO, see pages 2–3.

We'll close this chapter with some cold, hard truth. Although strategic writing is the ideal for successful organizations, it's not always the practice. Some organizations lack leaders who understand the importance of strategic communication; other organizations become distracted and overwhelmed by the many demands of successful management; still others may lack writers with your knowledge and skills. The authors of this book often receive documents from our graduates. Sometimes a former student will include a note that says something like this: "I know this is off-strategy, but the client insisted" or "This seems off-strategy to me, but my boss made me do it this way because it's how we've always done it."

As young professionals, we encountered the same problems. We've been there, and we sympathize. Our best advice is to try some strategic communication with your boss and your clients. Surely they want to succeed. If you can politely emphasize the advantages of strategic writing (without sounding like a pushy know-it-all), you may succeed in transforming your organization's communications.

On-strategy: That's our message.

# 1B The Importance of Good Writing

The authors of this book have lost count of the dozens and dozens of employers (your future bosses) we've interviewed. And those professionals all say the same thing: "Please give me someone who can *write*!" Yes, employers want diplomatic team players. And they want hard-working graduates with social media experience. But they almost always begin by asking for good writers.

We've noticed a similar trend in surveys of employers that list necessary job skills for success in strategic communications professions such as public relations. Good, strategic, multimedia writing usually tops the list.

This book is all about helping you become a first-rate strategic writer. As you already know, strategic writing involves delivering goal-oriented messages—messages that are on-strategy. Strategic writing also involves carefully crafted sentences. You don't want to distract your audiences with bad grammar, sloppiness or wordiness. The appendices of this book contain guidelines on punctuation, grammar, style, editing and proofreading. We hope you'll review those. What follows here are tips for strengthening sentences.

## 10 TIPS FOR WRITING BETTER SENTENCES

1. **Challenge *to be* verbs**: Challenge every appearance of *am, is, was, were, be, being, been* and every other form of the *to be* infinitive. Sometimes a *to be* verb best suits the needs of a sentence, but often you can find a stronger, more evocative verb.

| **Original** | **Revision** |
|---|---|
| He *will be* a good communicator. | He *will communicate* well. |
| We *are inviting* you. . . | We *invite* you. . . |

2. **Use active voice**: By active voice, we mean active subject. In active voice, the sentence's subject does the action described by the verb. In passive voice, the subject doesn't do the action.

*continued*

| **Passive Voice** | **Active Voice** |
|---|---|
| Our profits *were affected* by a sales slump. | A sales slump *affected* our profits. |

Passive voice is grammatically correct, and it's the right choice when the action is more important than the action's doer (for example, "She was fired"). But passive voice can seem timid, and it requires a weak *to be* verb. In contrast, active voice is confident and concise.

3. **Challenge modifiers**: Modifiers (adjectives and adverbs) can strengthen a sentence by sharpening your meaning. But sometimes they prop up poorly chosen words, especially imprecise nouns, verbs and adjectives. A precise, well-chosen word needs no modification.

| **Original** | **Revision** |
|---|---|
| We are *very happy*. | We are *ecstatic*. |
| *Quickly take* your report to the client. | *Rush* your report to the client. |
| He is *rather tired*. | He is *tired*. |
| Please deliver the package to our *headquarters building*. | Please deliver the package to our *headquarters*. |

4. **Challenge long words**: If a long word or phrase is the best choice, use it. Otherwise, use a shorter option.

| **Original** | **Revision** |
|---|---|
| Utilize | Use |
| Prioritize | Rank |

5. **Challenge prepositional phrases**: To tighten sentences, turn prepositional phrases into shorter adjectives when possible. Avoid a string of prepositional phrases.

| **Original** | **Revision** |
|---|---|
| I will present the report in the meeting *on Thursday*. | I will present the report in *Thursday's* meeting. |
| We will meet *on* Thursday *in* Weslaco *at* the Lancaster Hotel *on* McDaniel Street *near* the park. | We will meet Thursday at Weslaco's Lancaster Hotel, 1423 McDaniel St. |

6. **Challenge long sentences**: How long should a sentence be? Long enough to make its point clearly and gracefully—and no longer. Challenge sentences that have more than 25 words; realize, however, that some good sentences will exceed that length. As discussed earlier, you can tighten sentences by eliminating *to be* verbs, modifiers and prepositional phrases.

*continued*

7. **Avoid overused expressions**: Clichés such as "It has come to my attention" and "I regret to inform you" lack original thought. They sound insincere. Overused figures of speech such as "He's a fish out of water" don't create the engaging image they once did. Overused expressions suggest to readers that you didn't take the time to devote clear, serious thought to the message you're sending.
8. **Avoid placing important words or phrases in the middle of a sentence**: The beginning of a sentence breaks a silence and calls attention to itself. The last words of a sentence echo into a brief silence and gain emphasis. The middle of a sentence generally draws the least attention. A writer friend of ours says, "Words go to the middle to die." (This dead zone can be useful for minimizing the impact of bad news.)
9. **Keep the focus on the reader**: Tell readers what they want and need to know—not just what you want them to know. Keep the focus on how they benefit from reading your document. Talk to them about themselves and what your message means to them.
10. **Read your sentences aloud**: At least whisper them to yourself. That's the surest way to check for effective sentence rhythms. Reading aloud also can be an effective editing technique.

Good writing is also concise, so we'll end this chapter.

# 1c Research, Planning and the Writing Process

Good writing is more than just good luck and natural talent: It's the result of a logical process. Because the writing process can seem intimidating (or just plain hard), some people prefer to just rush in and start writing. But that's like leaving for a Spring Break trip with no destination, no map, no budget—and no hope. Other writers may feel so overwhelmed that they avoid the job until it's too late for their best work.

Good writing isn't easy. There's nothing wrong with you if you find writing to be hard work. You can, however, make that hard work a little easier by following a nine-step writing process.

## Step One: Research

This book shows you how to write dozens of documents for public relations, advertising, business communications and sales and marketing. For each document, we begin with an analysis of purpose, audience and media. We recommend that you do the same.

Begin your research by defining the document's purpose: What is its goal? What should it accomplish? What business goal does it support? With your answers to these questions, you should begin to answer another purpose-related question: What should be the one, key strategic message of this document?

Now extend your research to the target audience of the document. To whom are you writing? Audience research generally falls into three categories: demographic, psychographic and behavioral data. Demographic data consist of nonattitudinal facts such as age, income, gender, educational level, race and so on. Psychographic information contains attitudinal details about values, beliefs, opinions and, of course, attitudes. Psychographic information can include political and religious beliefs, personal ethics codes, goals in life and so on. Behavioral information describes habits and actions, such as media usage. Use your research to deeply understand your readers. Perhaps the most important question you can answer is why members of your target audience should care about your document. What's in it for them?

With your understanding of your target audience, you might want to refine the one, key strategic message you've begun to identify.

Finally, you should gather information about the medium or media you'll be using. Will you use Snapchat? Digital ads? Special events? Mobile messaging? All of the above? The characteristics of your chosen media can help you further refine your one, key strategic message. One of the best ways to select the best media for your message is to study your target audience. Which media does it prefer in this situation?

## Step Two: Creativity/Brainstorming

Some documents, such as advertisements, newsletter features and proposals, call for a high degree of creativity. Other documents, such as news releases and business reports, are more straightforward. When your one, key strategic message requires creativity, consider using a basic five-step approach to developing ideas. Advertising expert James Webb Young has written that the creative process consists of these steps:

1. Gathering research
2. Thinking about your research
3. Concentrating on other matters and letting your subconscious mix your research with other things you know (such as history, music, literature and movies). Young believed that a new idea was really a combination of two other ideas, facts or themes.
4. Recognizing when your subconscious reports back a great idea
5. Refining the great idea

A process known as brainstorming can assist the creative process. Brainstorming usually is a group activity in a comfortable setting. Group members toss ideas back and forth, building on one another's ideas, reviewing key research findings and encouraging everyone to be innovative. Brainstorming works best when two rules apply: No one's idea gets ridiculed, and no one worries about who gets the credit.

## Step Three: Organizing/Outlining

You've gathered all the necessary information. You've identified a key message and, perhaps, developed a creative approach. Now it's time to determine what to include and how to organize that information.

Many things affect organization, including the target audience's interests, the type of document you're writing and the importance of each piece of information. The best general guidelines for good organization are to consider your audience (what order of information will keep it interested?) and to be logical. You should have a reason for the order of presentation: One part of the document should lead logically to the next.

Writing an outline, whether it's formal with Roman numerals or just notes scribbled on a restaurant napkin, will help you refine and remember your document's organization.

Don't be surprised if you change or reorganize items as you write. New options may appear as you progress. (Experienced writers sometimes can create outlines in their heads—or, as they begin to type, they type a few organizational ideas and then begin composing.)

## Step Four: Writing

Finally. Now for perhaps the hardest part of the writing process. Again, writing is tough work for most of us. If you just can't get the first few sentences, start somewhere else. Your outline allows you to do that. And don't worry about getting the words just right in your first draft. It's more important to get the ideas and meanings right.

## Step Five: Revising

One truism about writing says, "Good writing isn't written; it's rewritten." Even if you love your first draft, set it aside for as long as possible. Return to it fresh, and be critical. Poet and novelist Robert Graves recommended imagining that your intended reader is looking over your shoulder and saying, "But what does that mean? Can't it be clearer? What's in this for me? How do I benefit by reading this?"

You might also try reading your document aloud. This can be a good way to catch mistakes or language that doesn't flow well.

Writers who get serious about revision sometimes find that they have accidentally memorized all or parts of a document. With the document temporarily lodged in their memories, the writers are able to revise it as they eat lunch, ride in an elevator or drive home. This may sound excessive (even weird), but it illustrates the point that good, successful strategic writers don't settle for first drafts.

## Step Six: Editing

Sure, colleagues may edit your document. But you should be the document's first editor. Think of editing as the last fine-tuning before you send the document to your boss. Editing consists of two parts—macroediting and microediting—and you should do both. Macroediting involves looking at the "big picture" of the document. Is the document's key message clear and goal-related? Does the document appeal to readers' self-interests? Does it cover the important parts of who, what, when, where, why and how? Is it well-organized—does one section lead logically to the next? Is the format—the way it looks on the page or screen—correct?

Macroediting also can involve a final revision. Can you find a precise noun to replace a current adjective–noun combination? Can you find a precise verb to replace a current adverb–verb combination? Are you using boring *to be* verbs too often? Can you find more interesting action verbs?

Microediting is proofreading. It involves going through the document one sentence at a time and double-checking grammar (including spelling and punctuation) and accuracy. *Double-check all names, dates, prices and other facts.* Use your computer's spellcheck program, but don't rely on it exclusively. Use a dictionary to look up every word or phrase that could be wrong. Double-check the accuracy of quotations. Microediting is best done backward, starting with the document's last sentence. Moving backward breaks up the flow of the too-familiar document. Moving backward makes the document sound new and different; it helps you focus on each sentence. You'll see what you actually wrote instead of what you meant to write.

## Step Seven: Seeking Approval

What could be hard about this stage? All you do is give the document to your boss and anyone else who needs to approve it before distribution. But experienced writers know that this can be one of the toughest steps in the writing process. You've done your best with the document, and you're committed to your approach. What if someone with authority wants to change part of it—or all of it?

Keep an open mind. Would the proposed changes make the document more strategic? That is, would the changes help it reach its goal more effectively? If so, swallow your pride and realize that a successful document often requires a team effort. But if the proposed changes seem to hurt the document's strategic value, do your best to politely debate the revision. Keep everyone's attention focused on the goal.

Never send a document to the target audience without undertaking this approval stage. By this point in the writing process, you're probably too close to your document. It's hard for you to be objective. The document now needs other reviewers and editors. And that can be hard. Avoid being a *prima donna*—that's the term given to temperamental opera singers who won't accept advice because they think they're perfect.

## Step Eight: Distributing

You must now send your document out into the world—or at least to the target audience. You may not be responsible for distribution, but you have a major investment in the document's success. Be sure you know where it's going and how it's getting there. And then be sure that it got there. As we said earlier, the best way to deliver a document is whatever way the target audience prefers. Be sure your research includes *how* the target audience wants to receive the information.

## Step Nine: Evaluating

In one sense, you began to evaluate your document much earlier in the writing process. When you considered different creative approaches and when you revised and edited, you were evaluating. In the approval stage when others edited your document, they were evaluating.

But now it's time for the big evaluation: Did your document succeed? Did it accomplish its strategic mission and fulfill its purpose? Learning the answers to these questions can help you do an even better job next time. If your document succeeded, why? If it failed, why? Did it have the desired effect on the target audience? Was its distribution effective and efficient?

Because strategic writers are so busy, evaluation can get overlooked in the rush to the next assignment. However, evaluation of past documents can lead to future successes.

The top three problems your authors see in student writing are a lack of research, a lack of strategic (goal-oriented) focus and a lack of polish (too many first drafts with small errors and awkward passages). We know that the writing process recommended here can seem like busywork. It can seem like something that textbook authors write just to fill pages or professors say just to fill class time. If you're doubtful about the writing process, we ask you to try it before rejecting it. We think the experience will make you a believer—and a better writer.

# 1D Writing for the Web

No other audience in strategic writing is as potentially diverse as a web audience. Your viewer could be the person next door or someone in Algeria. The web may be a mass medium, but it is still one-to-one communication. After all, people don't surf the web in groups.

People also don't read websites in the same way they read print: A study by Jakob Nielsen, a web usability researcher, found that almost 80% of users scan webpages rather than read them word-for-word. His research concluded that people read information on a computer screen 25% slower than information in print. This, coupled with the fact that people must first find your website before they can read it, means that writing for the web requires a different approach from writing for print. For this medium, it's essential that you hone your skills as a strategic writer. Your job as a web writer is to first make your site easy to find and then hook the user with beneficial content before they impatiently click away.

A website is an active, nonlinear medium. That means that the user doesn't necessarily start at the top and read to the bottom. A website, like other social media, is user-driven; the user decides when, how much and where to read next. Typically, web users are on a specific mission, looking for answers to an immediate problem. They want to sort through content, make a decision and then act. They are not interested in verbose explanations or long narratives. Thus, concise, tight writing is essential.

One main difference between print and online documents is that the former are tactile: Readers can hold a book or report in their hands and flip through the pages. That's not true on the web unless the user prints out the entire site. Users entering a site have no idea how many pages it includes or exactly what information it contains. That's why good navigation is critical. As a web writer, you must always provide cues as to where the user is in the site and how they can navigate to additional information. Navigation clues include page titles, subheadings, description tags and links.

## Search Engine Optimization

Typically, the user will enter your site through a search engine query. Therefore, your first job as a web writer is to write so users can find your site.

When a user searches for information via a search engine, the engine returns a list of webpages that match the query, placing the best matches first. Thus, your goal as a web writer is to get listed near the top of search responses. You do this by creating webpages that search engine crawlers will determine are matches for keyword searches.

### Top Search Engines in the World (2020)

Google, Bing, Yahoo!, Baidu (China), Yandex (Russia), DuckDuckGo, Ask.com, AOL.com, WolframAlpha and Internet Archive

A **keyword** is a word or phrase included in the website content to help a search engine locate particular pages. Users also type keywords into search engines seeking relevant websites.

Thus, **search engine optimization (SEO)** plays a vital role in getting your website found. SEO involves editing content and inserting HTML coding to increase the relevance of specific keywords and to remove barriers to search engine crawlers. Theoretically, the better optimized your website, the higher on the search results page it will appear.

SEO is not a science or an art, although many in the field would argue it is a bit of both. Because search engine software is proprietary, Google, Yahoo!, Bing and the others don't want you to know exactly how they sift through data. In fact, the processes are ever-changing, so it's impossible to know exactly what formula will work best at any given time. Google reportedly changes its algorithm 500 to 600 times per year and says it uses hundreds of different ranking factors that range from location of keywords to basic content reading level and the inclusion of videos. The search engines themselves do provide suggested practices that will help your site be found, ranked and indexed.

## Content and Keywords

So what's the best way to optimize your website? Google answers this question clearly in its guidelines: "Create a useful, information-rich site and write pages that clearly and accurately describe your content" (support.google.com). In other words, "Content Is King," as Bill Gates declared in 1996. The primary way to increase your rankings in Google, or any other search engine, is to:

- Provide original information not available elsewhere on the web.
- Provide information of value that can benefit users.
- Clearly describe the content of your website.
- Ensure that other sites link to your site. (If others reference your site, Google sees it as more reliable.)

- Use good navigation tools and structure in building your site. (For example, eliminate broken links, slow downloads and duplicate copy.)

> Google provides a detailed Search Engine Optimization Starter Guide that walks you through the best industry practices.

As discussed earlier, people find your website by typing keywords into a search query. Your task as a web writer is to use keywords in the content of your website that match users' search queries. The higher up your company or organization appears on the search response, the more likely a user will click through to view your website.

Website owners often think they know the viable keywords that people will use to find their site. This isn't necessarily true. Many site owners aren't objective about their site's content and are too tied to industry jargon. The following suggestions can help you identify keywords for your site.

- **Create a list of potential keywords**: Ask yourself: What other names or terms are related to your product or service? What is your site hoping to do or promote? What services will it provide?
- **Brainstorm**: Ask friends, colleagues and loyal customers or users for potential keywords.
- **Visit discussion forums**: Record the words prospective customers are using. Think like a consumer. What are they searching for? What is their intent?
- **Know your competitors**: What keywords are they using?
- **Use single words as well as phrases**: Remember the term "keyword" doesn't refer only to single words but also to phrases, called long-tail keywords.
- **Avoid industry jargon**: Users might not understand trade-speak.
- **Use a keyword research tool**: Google Adword Keyword Planner or the Bing Keyword Research Tool can identify possibly overlooked keywords.
- **Dig deeper**: Use the keyword research tool to learn more about each word or phrase, including its conversion rate (its success in drawing viewers), search volume (how often the word has been entered into search engines) and competition (similar keywords).
- **Narrow your choices**: Refine your list of keywords to include a mix of broad and highly targeted words. Also create some "negative" keywords—ones that do not describe your site's product or service. Obviously, avoid using these.
- **Finalize your list to 15–25 keywords**: Consider ranking them as primary or secondary keywords.

Now that you have your list of keywords, how do you use them to write strategically?

- **Content**: Use three to four keywords per page when writing your content. Using more will dilute the content and could be viewed as "keyword stuffing." Google no longer gives good rankings to pages that employ this technique.
- **Heads, Links and Navigation**: Use keywords in page content, page titles, internal headlines, <Alt text> (text that substitutes for images that a browser can't open) and links, including navigation links. Vary the titles of different pages in order to use as many keywords as possible.
- **Meta Description**: Use keywords in the meta description for the page, a sentence of up to 155 characters that describes page content. Description tags are HTML-coded passages that don't appear on the visible page. Search engines often use this embedded tag as the snippet that appears along with the search results. If you don't provide a tag, the search engine will construct one from the contents of the page.
- **Ranking**: PageRank was developed by Larry Page and Sergey Brin, Google's founders, as a system for ranking webpages by giving each page a relative score of importance. While the exact formula of PageRank still remains a secret, quality backlinks (incoming links) and efficient internal links seem to increase search engine results.

As the needs of your site and its visitors change, so should your keywords. Constantly monitor your keywords, and update them as needed. Remember: Keywords are ultimately about people—your audience. The better you know your target audience, the more targeted your keywords will be. The best advice still remains: Write information-rich content that is relevant to your audience.

## Web Writing That Works

> Vigorous writing is concise. A sentence should contain no unnecessary words, a paragraph no unnecessary sentences, for the same reason that a drawing should have no unnecessary lines and a machine no unnecessary parts.
>
> William Strunk Jr., in *The Elements of Style*

A modern-day web writer can benefit greatly from the advice William Strunk Jr. penned in 1918 in his "little book" of essential rules for good grammar and composition. Less is more when it comes to web writing.

- Documents intended for online reading should rarely be longer than 1,000 words. A good target is 500 words or less. If your website is most likely to be viewed on a mobile device, a good target length might be closer to 250–300 words per page, according to Google. Remember: People find it more difficult to read information on a computer screen or hand-held device, and thus read it more slowly. Edit. Then edit again. Use the 10 Tips for Writing Better Sentences on pages 11–13.

- Split content into information bytes. To attract and hold a reader, make text short with one idea per paragraph. Strive for paragraphs no longer than 50 words and sentences that contain no more than 15 words. Use short, simple words. One of the handiest tools for the online writer is your software's word-count function. Use it!
- Skip introductions or welcomes. Instead, establish why this page is important in a single, short statement. In other words, start with your conclusion! Give a concise overview of the information first.
- Use the inverted pyramid organizational scheme (page 100): In passages of more than one paragraph, information should become progressively less important.
- Put key information in the first sentence of a paragraph. In fact, the first three words should be loaded with information.

  **Don't Write**: 2021 was a big year at Riverview Community Medical Center with surgeries for knee replacements reaching all-time highs.

  **Do Write**: Knee replacement surgery soared at Riverview Med in 2021.
- Use links to provide more in-depth coverage or background information. Think in terms of layers. The top layer is the outer skin of your website, and as you go deeper inside, you reach the meat. Use internal links to appeal to specific audiences.

  **Don't Write**: For articles about knee replacement procedures Click Here.

  **Do Write**: Advances in knee replacement procedures at Harvard show. . .
- For in-depth information, use the bite-snack-meal approach developed by Marilynne Rudick and Leslie O'Flahavan. On the website homepage a headline serves as the bite and is a link to the full article (the meal), which appears on an interior page. The snack is a two- or three-sentence summary of the article beneath the headline. This allows readers to choose what level of detail they want.
- Be objective. Don't use a promotional writing style. Web users want straightforward facts. Link to other websites to increase credibility.
- Make text timeless (avoid words such as *recently* and *today*).
- Use sentence fragments where helpful. Print documents generally call for complete sentences, but complete sentences often aren't necessary in online writing. Sentence fragments often let you pull information-carrying keywords to the front.

  **Do Write**: *25% less fat!*
- Use numerals rather than spelling out numbers, even at the beginning of a sentence or in a headline. This doesn't apply to really large numbers, such as a billion.

  **Do Write**: 10 Reasons to Choose Riverview Med

- Use bulleted lists to make text scannable. Bullets (•) slow down the eye and highlight information. Use bullets when the sequence doesn't matter; use numbers when it does. In websites, bulleted information should be one or two words in length.
- Make text scannable by using headings, subheadings, links, keywords and lists.
- Highlight text for emphasis. Put keywords (not whole sentences) in bold or colored text. Avoid blue and purple, however. The default settings on most computers are blue for unvisited sites and purple for visited ones.
- Limit the use of italics or underlining for emphasis. Italic type is more difficult to read on a computer screen (although advances in vector graphics and higher-resolution monitors have minimized this issue), and underlining may be confused with links. Think in terms of eyebytes—two or maybe three emphasized words. Use about three times as many as you would in a similar print document.
- Integrate graphics within the text. Use charts and visuals to explain complicated information. Avoid graphics with movements that might distract the reader from the content. Caption all graphics clearly. Clear captions also assist visually impaired viewers whose specialized computers may not display the graphics.
- Make printer-friendly versions (PDFs) for information likely to be printed.

## Web Headlines

Website headlines must work overtime. Their job is to help users skim through copy and locate their desired information. Headlines are probably the single most important words on your webpage. After all, they have to appeal to three audiences: your Reader, Search Engines and Social Media. It's a rare case when the same headline can captivate all three, however.

Your first consideration should always be the readers. If your headline doesn't immediately grab readers' attention, they may click through to another site, and you've lost them. Search engines seek very specific content connected to an intended search, so your headline must contain those keywords.

- Write headlines, subheadings and internal headlines that summarize content. Since users can enter the site from any page and freely move among pages as they choose, make each page stand on its own and not be dependent on previous pages. Headlines often appear separated from their associated story in search results, tweets, blog posts or newsfeeds. That means they must be meaningful out of context.
- Make every word in a headline meaningful. Don't be cute. Don't be promotional. Don't run the risk of confusing international audiences with slang. The headline should clearly explain what information follows.

  **Don't Write**: Get the Most Bang for Your Buck

  **Do Write**: Increase Your Sales by 15%

- Write short, direct headlines that feature keywords. Try to write headlines that average about five to eight words. Google will only show the first 55–60 characters on a page title. Pay attention to what words might be cut off. Remember short headlines are better for tweets, too.
- Front-load your headlines, placing the most important keywords first.

  **Don't Write**: Scientists Reveal Slower Metabolism Causes Weight Gain

  **Do Write**: Weight Gain Due to Slow Metabolism, Study Says
- Use passive voice when it allows you to place keywords first in a headline or sentence.

  **Active voice**: *flavours!* Introduces Soy Milk Diet Bars

  **Passive voice**: Soy Milk Diet Bars Introduced by *flavours!*
- Use numerals rather than spelling out numbers.

  **Don't Write**: Five Tips to Weight Loss

  **Do Write**: 5 Easy Tips to Weight Loss
- Use lists and how-to titles.

  **Do Write**: How to Lose 10 Pounds This Week
- Use internal headlines. The job of an internal headline is to select and highlight a word, phrase or idea from the following few paragraphs that will make the reader want to keep reading. Internal headlines also serve to break up uniform blocks of type into manageable chunks.

Finally, pay attention to websites that you visit. What works? And just as important, what doesn't?

# 1E Writing for Social Media

Social media may be as commonplace as eating lunch or brushing teeth, but they're still new and ever-changing tools. For people born at the turn of the century, most of these media were created after they first attended preschool. And the evolution of social media continues at the speed of imagination.

Let's start this discussion of social media with some basic definitions:

- **Social media**: Software, platforms and websites that allow people to share information
- **Social listening**: The practice of identifying engagement opportunities by monitoring social media conversations that align with our personal and organizational interests
- **Social engagement**: Interactions that can result from social listening and joining the conversations

These definitions suggest that the first step in the strategic use of social media is listening—not unlike research being the first step of strategic writing. Understanding your organization's target audiences—both present and potential—is key to successful strategic communication. These definitions also suggest the second step in strategic writing: planning. Organizations need to know what they want to accomplish through social media, whether it be growing and empowering a network of passionate supporters or providing a human face for the organization through shared stories.

Social media posts are an increasingly important part of strategic writing. In one recent survey, international CEOs identified social media as their company's most valuable communications channels. In the same survey, international public relations professionals rated social media as growing in importance faster than other communications channels.

Given that importance, how often should an organization post? Social media experts surveyed by *Inc.* magazine recommend these posting frequencies:

- Twitter: 10–15 tweets per day
- Instagram: 1–2 posts per day
- Facebook: 1 post per day

- LinkedIn: 1 post per day
- Pinterest: 10 pins per day

To this advice, your authors would add two things:

- Each post should communicate something of interest to your target audiences consistent with your goals and values.
- The above numbers indicate original posts. Those numbers don't govern how often you should join conversations or respond to inquiries delivered via social media.

Because social media platforms evolve rapidly and differ from one another—Facebook isn't Twitter, which isn't Snapchat—it can be tough to supply tips that apply to strategic writing for *all* social media. But experts agree that some basic guidelines do exist.

## 10 TIPS FOR SOCIAL MEDIA WRITING

1. **Pause to plan**: Don't write until you've considered purpose, audience and media. The purpose of social media generally is to encourage audience engagement and interaction with your organization. Your audience is technologically literate, and the media should be those preferred by your target audience.
2. **Focus on audience wants, needs or interests**: Don't post unless you're addressing one of those categories.
3. **Create a social media personality (voice) for your organization**: Experts recommend informality with a slightly edgy sense of humor—but that advice is not appropriate for all organizations, such as nonprofit groups that address serious health issues, or in all circumstances. Even humorous personalities sometimes must deliver serious, straightforward messages.
4. **Don't shoot from the hip**: Do be proactive with posts, and respond quickly to opportunities and to the concerns of key audiences—but be sure that every social media message is reviewed, edited and approved before posting. Social media dashboards (page 247) that require levels of authorization can assist with the approval process.
5. **Stay up to date**: Social media dashboard technologies allow you to schedule social media messages; coordinate those messages with information from your databases that store information on customers and/or other key audiences; and organize the editing and approval process.
6. **Engage your readers**: Again, create posts that captivate readers and, ideally, become individual relationship-building conversations. Strategic writers often craft social media messages to pull readers to a so-called landing page within their organization's website.

*continued*

7. **Be concise**: Tweets are limited to 280 characters and spaces, and experts at Sprout Social recommend aiming at 100 characters for such posts. Facebook posts of 40–80 characters spark the highest levels of engagement.
8. **Use visuals**: Posts with photos, videos or GIFs draw more attention and engagement from readers than do simple text messages.
9. **Use analytics to learn which posts work best with which audiences**: Facebook, for example, has a Page Insights section, just as Twitter has Twitter Analytics. (If we're not studying reactions to our messages, we don't know if we're succeeding.)
10. **Think beyond "buy now" messages**: The Content Marketing Institute recommends restricting direct sales messages to just 20% of your social media posts. The remainder should be entertaining and informative messages, including so-called hacks and how-to posts and responses to current events.

Perhaps the greatest challenge in the strategic use of social media is finding a correct balance in your organization's social engagement. While there are many positives in encouraging interactions with your various target audiences, there is also risk. Poorly planned posts can attract worldwide scorn, and even private posts by employees can reflect poorly upon an organization. It's imperative, therefore, that organizations establish social media policies, as well as train employees in their use. And not all threats to social media success come from within. Again, organizations should regularly monitor the social media environment for potential threats as well as opportunities.

One more piece of advice: Be a student of social media writing. Learn from what works—and what doesn't.

# 1F Writing for the Ears

Have you ever noticed that it is often easier to silently read a book or email than it is to read it aloud? You may not have realized it at the time, but that was your first experience with the difference between writing for the eyes and writing for the ears.

Aural writing—writing for the ears—is increasingly important in the age of digital and social media. It is the style of writing used when making a live presentation in front of an audience (such as a speech or sales presentation) or for one that is recorded or transmitted to the audience (such as radio/television commercials, online videos or podcasts).

There are three major similarities between print-style writing (which is created for the eyes) and aural writing (which is created for the ears—*and* eyes, in the case of video). We write each in a manner best suited to the audience. We write each in a manner best suited to the purpose behind the message. And we write each in a manner best suited to the medium used to convey the message.

The major difference is that writing for the ears uses language and formats that make it easier for the narrator to read the copy and for the listener to understand it. Aural communication, what we hear, is linear. That means there are no second chances. When we're reading, we can pause, reflect and re-read a sentence. However, that doesn't happen when listening to the spoken word. Once the message is delivered, it's gone—unless we're able to rewind a recorded message, which is more difficult than simply re-reading a written sentence. For this reason, the aural style of writing features short, active voice, subject-verb-object sentences with key information at the start of a sentence. The style remains the same regardless of the message.

Writing for the ears is different from writing for print in several other ways. Print media are better suited for details. An audio listener or video viewer can easily get lost in an avalanche of facts and figures. That's why the use of broad concepts, tangible examples and big ideas is preferred in aural writing. It's also why broadcast and podcast writers repeat key phrases and names—especially in advertising and other persuasive messages created to be remembered.

And, of course, major differences separate writing for audio, which has no pictures, and writing for video, which is dominated by pictures. In video, words and pictures must work in unison. For example, strategic writing for television commercials and video news releases involves designing images as well as crafting the words that enhance them.

The value of learning how to write for the ear has grown with the increasing prominence of social media videos as well as podcasts—a digital audio file downloaded from the internet for use by the listener on a device of their choosing. Podcasts have become important in the branding of organizations. They also help create a connection, a relationship, between the source and the listener. Podcasts also present their sources as trusted experts.

Consider *Serial,* a 12-part series of podcasts produced and distributed by National Public Radio in 2014. It was an evolving story, a murder mystery, about the 1999 death of a young woman in Baltimore and the questionable conviction of her boyfriend. *Serial* became one of the most-downloaded podcasts in history. And what did NPR get for its efforts? It attracted a younger, loyal group of subscribers who may assist in the nonprofit organization's future fundraising. Because of its success, *Serial* was extended for new stories in subsequent seasons.

## 10 BASIC RULES OF WRITING FOR THE EARS

1. **The announcer has to breathe**: Stick to short sentences of 20 or fewer words. The shorter, the better.
2. **One at a time**: Only one major idea per sentence. Avoid compound sentences. (The word *and* should raise a red flag.)
3. **Make it personal**: It really doesn't matter how many people are listening. In reality, you are speaking to one person at a time. Speak directly to that person. The best copy is conversational, so write like you talk. Sentence fragments are acceptable, just as long as they make sense.
4. **First things first**: In contrast to writing for print media, attribution of paraphrased quotations should be at the beginning of the sentence. This makes it easier for the listener to distinguish opinion from fact. All titles go before a person's name. That goes for official titles, such as "Mayor Mary Smith," and unofficial titles, such as "community activist Mary Smith."
5. **Write S-V-O**: Use simple subject-verb-object sentence structures. Eliminate *to be* verbs whenever possible.
6. **Use active voice**: Make the subject the doer of the action. "Lincoln wrote the Gettysburg Address" is better than "The Gettysburg Address was written by Lincoln."
7. **There's nothing like the present**: Use present tense—except when past tense is necessary. Electronic media are instantaneous. Present tense expresses this sense of urgency. It is especially important that attribution be in present tense, preferably using the neutral *say* and *says*.
8. **Write it as you would say it**: Avoid bureaucratic jargon. Speak the language of your audience. When using initials in lieu of an organization's name, use hyphens between the letters if the announcer is expected to pronounce each

letter. (For example, "F-B-I" for the Federal Bureau of Investigation. However, the common second-reference pronunciation for Mothers Against Drunk Driving is "MADD," pronounced as a word rather than four separate letters.)

9. **Know your numbers**: In a broadcast script, write words for single-digit numbers (for example, "six" and "nine"). Use figures for two- and three-digit numbers (for example, "23" and "147"). For numbers with four or more digits, use a combination of figures and words (for example, "156-thousand"). Because aural communication is better suited for big ideas than details, round-off large numbers and fractions unless precision is required. For example, "more than 25-thousand" is better than "25,389." However, there is a big difference between earthquakes that measure five-point-one and five-point-six on the Richter Scale. And spell out *dollars* instead of using the dollar sign.
10. **Make it easy to find**: For online audio and video files, create captivating titles that draw attention. Podcast program notes should include relevant keywords and phrases easily identified by search engines.

The format used in writing audio and video scripts focuses on the needs of the announcer and/or actors. Each script serves as a roadmap for how to present the message. In addition to the words that will be read aloud, the script contains instructions for the use of music, sound effects or recorded voices. Video scripts also contain visual instructions. To make a script easier for the announcer or actors to follow, you should use large typefaces and double-space the lines.

Although it is not unusual for television and film writers to use single-column scripts in long-form productions such as a television drama or a documentary, a two-column script is more common. This book will go into greater detail about script formats in later discussions of specific documents, including video and audio advertisements (see pages 203–209; 210–216).

One important detail of these scripts is the special language strategic writers use to communicate with producers, directors and editors. We'll close this discussion with a brief glossary of terms you should know to write scripts and talk the talk with the pros.

**Actuality**: A short snippet of a longer interview used in an audio script. Also known as a **Soundbite**.

**Chyron**: Words shown on a video screen. Also known as a **Super**. A slash (/) indicates a line break in a chyron message in a script.

**Crossfade**: The overlapping of audio or video as one source fades in and the other fades out.

**CU**: A close-up shot in a video script, often of a face, hands or feet.

**DOG**: Digital on-screen graphic. Also known as a **Watermark**.

**Dolly**: To physically move a video camera forward or backward rather than zooming, panning or tilting the camera from a fixed location.

**Establish, then under**: A description of playing music or a sound effect at full volume for a short time to attract attention or allow recognition, then lowering the volume to allow use under voices.

**Establishing shot**: In a video script, a wide shot (WS) that clarifies the scene for an upcoming sequence of shots.

**Fade**: In audio, a gradual decrease of volume. In video, a gradual darkening of a visible scene.

**MS**: A medium shot in a video script, often of a person shot from the waist up.

**Pan**: To move a camera's lens from left to right (or right to left) without moving the camera itself from a particular location.

**RT**: Running time. Specified at the end of audio and video production scripts.

**Sequence**: A group of related shots in a video script.

**SFX**: Sound effect or sound effects.

**Shot**: A still or moving picture taken from a particular location, width and angle.

**SOT**: Sound on tape. Refers to use of natural sound or music from a previous recording.

**Tilt**: To move a camera's lens up or down without moving the camera itself from a particular location.

**Under**: A description of background sound or music that runs unobtrusively beneath voices in an audio or video production.

**VO**: Voiceover. Words spoken by an unseen announcer.

**WS**: A wide shot in a video script, often of a building, a room or a group of people.

**Zoom**: The process of making images larger or smaller by adjusting the focal length of a camera's variable-length lens.

## TIPS FOR VISUAL STORYTELLING

Have you ever watched someone's home videos and thought they were similar to a toddler's first efforts with a camera? Shaky camera movements, rapid scene changes and an absence of a coherent storyline often characterize these homemade productions. However, that's OK. Most home videos serve their intended purpose: to remind us of people and places.

What works for a home video doesn't necessarily work in a strategic communication campaign designed to deliver a clear message to a targeted audience. Just like strategic messages told in words, strategic messages told in pictures have a specific structure designed to enhance the audience's understanding. Like all other strategic communication, visual storytelling involves research and planning. With that in mind, here are a few tips for visual storytelling:

1. **Decide where you are going**: Research your purpose and audience. Before you do anything else, determine the desired outcome. Even before you pick up the camera, you also need to understand any time, budget and logistical limitations.
2. **Put the words first**: Write the script first. It is more strategic to shoot pictures that match a script than to shoot first and write later. You often have only one opportunity to photograph a certain person or location. Writing first and shooting later ensures that you come away from your video session with all the material you need. A storyboard (page 212) also can help you plan each shot.
3. **Use a tripod**: Professional photographers use them because they help create a steady image. Remember that something that may seem like a small movement on a phone or tablet is magnified when it appears on a larger screen.
4. **Know your video grammar**: Think of each shot as if it were a sentence. Just as a paragraph is a series of related sentences, a sequence is a series of related shots. Just as a topic sentence expresses the main idea of the paragraph, an establishing shot sets the stage for the remaining shots within a sequence.
5. **Variety is the spice of life**: Vary the length, angle and width of shots within a sequence. Avoid predictable patterns that may cause the viewer to lose interest. Avoid putting two wide shots or two medium shots back-to-back. Know the desired length, angle and width of each shot—but also shoot each shot with different lengths, angles and widths to allow variety in editing and to provide just-in-case backup shots.
6. **Do not cross the axis**: The axis is the direction of action within a sequence. For example, when two people are in a shot, the axis is located on a line between them. All the shots within the sequence should stay on the same side of the axis. Otherwise, you confuse the viewer by reversing the flow of the action. It would be like showing a football player running down the field and switching to a camera on the other side; our player would suddenly be running in the opposite direction.
7. **Hit and run**: In visual storytelling, hit-and-run writing involves the relationship between pictures and words. Pictures and words are connected during an establishing shot in a sequence; that is, the words help explain the picture. During the remainder of the sequence, the relationship between pictures and words does not have to be as strong. But when the scene changes, the pictures and words should reconnect during the first shot of the new sequence (hence the name "hit-and-run").
8. **Continue action and setting**: It is possible to achieve the look of a multi-camera shoot using just one camera. By paying attention to detail and matching the

*continued*

action and setting of one shot with the next, even the most modest production can have that big-budget look.

9. **Avoid jump cuts**: Jump cuts are unnatural movements created by bad editing. They also are known as breaks in continuity. An example would be when a person sitting in a chair in a sequence's first shot is suddenly—a fraction of a second later—standing on the other side of the room. It is physically impossible and therefore appears unnatural.
10. **Cutaways are your friends**: Cutaways help maintain continuity by briefly shifting the viewer's attention to another subject. Using the previous example, it is easy to remove the jump cut—and the unnatural movement it creates—by placing a shot of another person in the room between the shots of the subject first sitting and then standing. It looks natural because the viewer rationalizes the action occurred off-camera during the cutaway.
11. **Empty frames are your friends, too**: This is another popular technique for maintaining continuity—especially when the same subject is in consecutive sequences. This is achieved by having the subject move out of the frame at the end of one sequence or into the frame at the beginning of the next. It looks more natural than having people pop from one location to the next in a fraction of a second.
12. **Let the action come to you**: It is usually best to have action come toward the camera rather than move away from it. The exception comes when movement away from the camera makes sense within the context of the production.
13. **Action is recorded, not created**: Avoid unnecessary camera movements and zooms. They tend to distract the viewer.
14. **Go easy on the special effects**: Use special effects only when they complement your message. Unnecessary effects distract the viewer.
15. **Use the right light**: Proper lighting gives the illusion of a third dimension in a two-dimensional medium. As with everything else, you want the subject to appear natural. Unnatural lighting will distract a viewer's attention. Avoid backlit situations, as well as those in which the lighting is diffuse/flat or creates a sharp contrast.
16. **Challenge authority**: One final piece of advice: It is OK to break any of these rules if doing so makes sense within the context of what you hope to achieve. However, you need to know the rules *before* you can break them.

# 1G Strategic Design

Words alone often won't suffice to convey your message as clearly and strategically as possible. That's why you need to understand design as it relates to strategic communication: Even if you don't see yourself as a designer, it's in your own best interests to understand basic design principles. Research shows that communication with a visual component is far more effective, persuasive and memorable. Whether you're producing an annual report, marketing proposal, website, brochure or print advertisement, good design principles will enhance the message. Good design attracts and holds the reader's attention. It amplifies the message and provides direction and order. Good design should work seamlessly with your words to reinforce the strategic message.

Graphic design uses typography, visual art, page layout, paper, computer software (design software) or other creative devices to combine words, symbols and images to enhance the overall message. Different typefaces create different moods and emotions. For example, a typeface appropriate for a wedding invitation would not work as well for a corporate annual report.

Wedding Invitation: *You Are Cordially Invited...*

Corporate Report: **Riverview 2023**

Adjusting the point size, line length, leading (line spacing), tracking (general space between letters) and kerning (the space between particular letters) has an impact on the overall appearance of the page.

Most often, graphic artists design pages using a grid system that provides an invisible structure for aligning and repeating elements on a page. Grids are meant to be flexible yet provide a continuity for a multipage project. A grid can help clarify design concepts such as balance.

The principles of design govern the relationship of the elements on a page or screen. While there are no absolute rules, less is usually more when it comes to design. The goal in good design is to create a harmony between elements that works in tandem with the overall message.

## Principles of Design

**Balance** means equalizing the weight on both sides of a centered vertical or horizontal axis. When elements spring out symmetrically from a central point, such as a point on an axis, they have radial balance; the central point provides a center of gravity for the design. Each element in a design has its own visual weight. For example, a photograph is visually heavier than a headline, which is heavier than body copy. The heavier the element, the more your eye is drawn to it. You can create symmetrical or asymmetrical balance. Symmetrical balance (Figure 1.1) centers elements along the vertical or horizontal axis and creates a more conservative, formal look. Asymmetrical balance (Figure 1.2) places elements off center and creates a sense of tension and movement. For example, a one-third/two-thirds division of space is more dramatic than dividing space in half along a centered axis. This Rule of Thirds divides your design into three rows and three columns. An intersection of these guidelines provides an ideal place to position your most important element. The Rule of Odds states that having an odd number of objects in your design is more interesting and, therefore, more pleasing to your audience than having an even number of objects.

**FIGURE 1.1**

**FIGURE 1.2**

**Movement** adds excitement and energy to your page by directing the path the eye follows (Figure 1.3). Movement can create unity through repetition and rhythm. For example, repeating a pattern of lines or shapes will force the eye to follow them. Movement can also be created through action. This can be done in the two-dimensional world by taking a "freeze frame" of an object in motion, such as a dancer or jogger. Additionally, the downward angle of the dancer's arm could lead the viewer's eyes to the next important element—a paragraph, for example. Good design directs the eye in the desired sequence of movement. Typically the eye enters the page at the top-left corner and moves across to the right, then diagonally down the page to the lower-left corner and then off the page in the lower-right corner—a basic Z pattern. For this reason, logos are frequently placed in the lower-right corners of print ads. Thus, a logo is the last thing on the page that the consumer sees.

**Emphasis** creates a point that acts as a bull's eye or focal point for the viewer (Figure 1.4). When a layout has no emphasis, nothing stands out, and the viewer doesn't know where to focus. Making an object larger, more sophisticated, more ornate, more brightly colored or closer to the foreground can increase its emphasis or dominance.

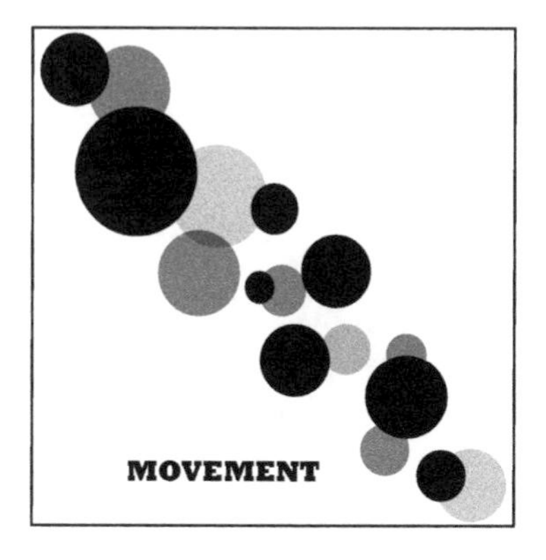


FIGURE 1.3

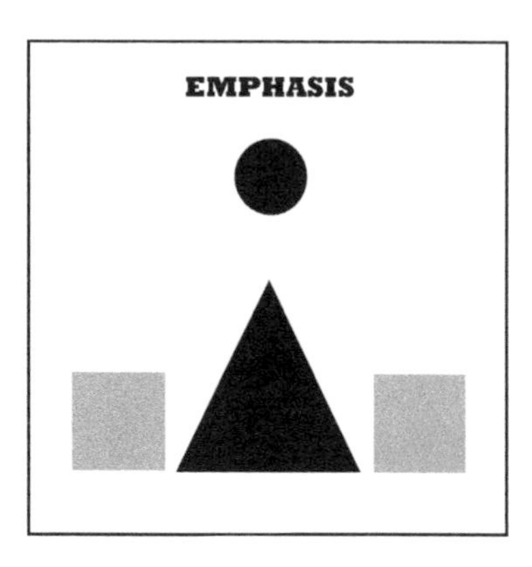


FIGURE 1.4

**Contrast** occurs when two related elements differ—the greater the difference between elements, the greater the contrast (Figure 1.5). Contrast adds variety to the total design and creates unity. Too much similarity of components in any design becomes monotonous; nothing stands out. Common ways to create contrast include establishing differences in size, shape, color, value (lightness or darkness), alignment (see *balance*), type, movement, direction or texture.

**Proportion** deals with how one element relates to another in terms of size, weight, shape, color or location (Figure 1.6). The dominant element is where the eye naturally goes first. Every layout should have one and only one dominant element. Elements that are placed closer to the center have less visual weight than elements in the corners. Good proportion creates harmony by having shapes fit properly in relationship to other shapes.

FIGURE 1.5

FIGURE 1.6

**Space** refers to the distance or area between, around, above, below and within shapes and forms. Positive space is the occupied area in a layout that contains copy, photographs, headlines or other design elements. Positive space dominates the eye and is the focal point of the layout. Conversely, negative space (also known as white space) comprises the unoccupied areas around the other elements. White space (Figure 1.7) helps balance the elements on the page and gives the eye a place to rest.

FIGURE 1.7

FIGURE 1.8

**Unity** brings order to your design. It makes all elements appear to belong and work together (Figure 1.8). Unity can be achieved by:

- Similarity: repeating colors, shapes, textures, values and related design elements
- Continuity: treating different elements in a similar manner
- Alignment: arranging shapes so that the edge of one element leads the viewer's eye to another element
- Proximity: grouping elements together so that several are viewed as one

**Color** is not generally considered one of the basic elements of design, but it plays an important part in communicating a message. The colors on the spectrum (the full range of colors within light) evoke different temperatures: Blue, green and the neutrals white, gray and silver are cool colors. Cool colors visually recede on a page. Warm colors, such as red, yellow and orange, however, create excitement. They appear larger than cool colors and overpower cool colors when used in equal amounts. Color has three distinct properties: hue, value and saturation.

The colors of the spectrum are called **hues**. A color wheel is a circular arrangement of colors organized in the order of the light spectrum. Traditional color wheels use three primary colors: red, yellow and blue. Separating them are three secondary colors: orange, green and purple. The colors that divide the primary and secondary colors are tertiary (third-level) colors. They are named for their parent colors, with the primary color first (for example, yellow-green).

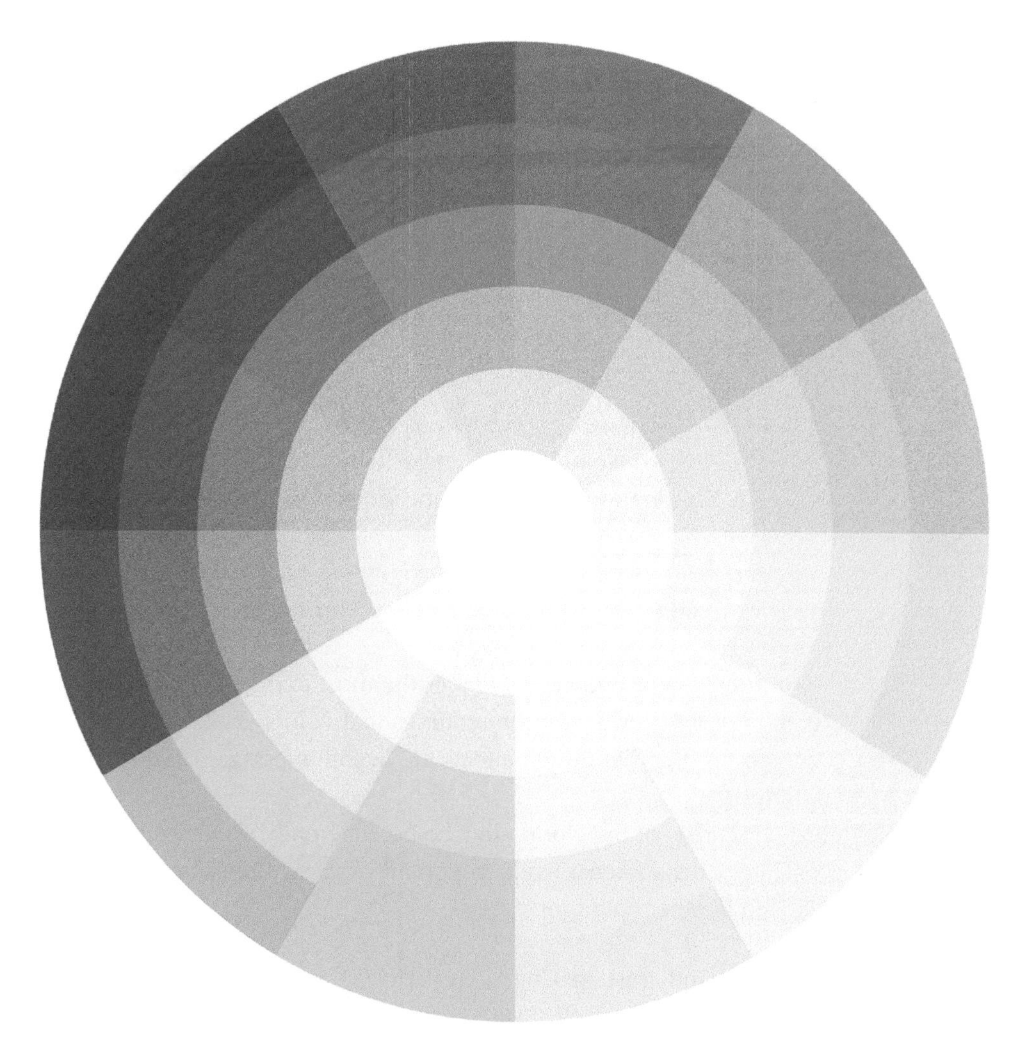

**FIGURE 1.9**
Credit: opicobello/shutterstock.com

The key to creating a successful color scheme lies in understanding the relationships between the hues on the color wheel (Figure 1.9).

- Monochromatic color schemes use tints and shades of only one hue. The hues can vary in value (lightness or darkness). This visual effect is extremely harmonious, quiet, restful and, depending on the range of values, subtle.
- Analogous color schemes comprise three colors side by side on the color wheel. They are easier on the eye and often have a very soft feel. This scheme can become a monochromatic effect with analogous accents, with one dominant and two additional colors.

- Complementary color schemes use two colors located opposite each other on the color wheel. Each pair contains one primary and one secondary color. When placed side by side, a complement enlivens the other color, making it appear brighter, stronger and more interesting to the eye. A split complementary scheme begins with one color and then adds the two colors on the sides of its complement.
- Triad color schemes contain three colors that are of equal distance apart on the color wheel. They may include tints or shades of a triad color. A triad creates a bold color scheme.

**Value** concerns the light and dark properties of color. The strongest contrast is black to white. Strong contrast is useful in directing attention. Variations in value can help create a focal point. Graduations of value can be used to create the illusion of depth.

**Intensity**, also called saturation, refers to the brightness of a color. A color is at full intensity when not mixed with black or white. You can make a color duller by adding gray.

Color is the visual component people remember the most and increases brand recognition by up to 80%. Artists have known for centuries that colors evoke emotions and strong feelings. How you use them in design affects the overall message. Colors create a *personality* for your brand.

Different cultures and areas of the world view colors differently, so choosing a color scheme for your design and its audience is an important process. Consider these colors and how they are viewed in other countries:

Black means high quality and trust in China. In Africa, it means age, maturity and masculinity.

White represents death, mourning and bad luck in China and Korea.

Blue means immortality in Eastern cultures and good health in Ukraine.

Green signals high-tech in Japan, luck in the Middle East, independence in Mexico and death in South America.

Yellow represents envy in Germany, mourning in Mexico and strength in Saudi Arabia.

Orange is considered auspicious and sacred in Hinduism. It represents sexuality and fertility in Colombia and love, humility and good health in Eastern cultures.

Red is good luck, joy and prosperity in Asian cultures, danger in Europe, purity and spirituality in India, mourning in the Ivory Coast and death in Turkey.

*(Source: Shutterstock, Inc.)*

Compare the meaning of colors noted here to the chart below (Figure 1.10) that shows generally agreed-upon color messages in Western society.

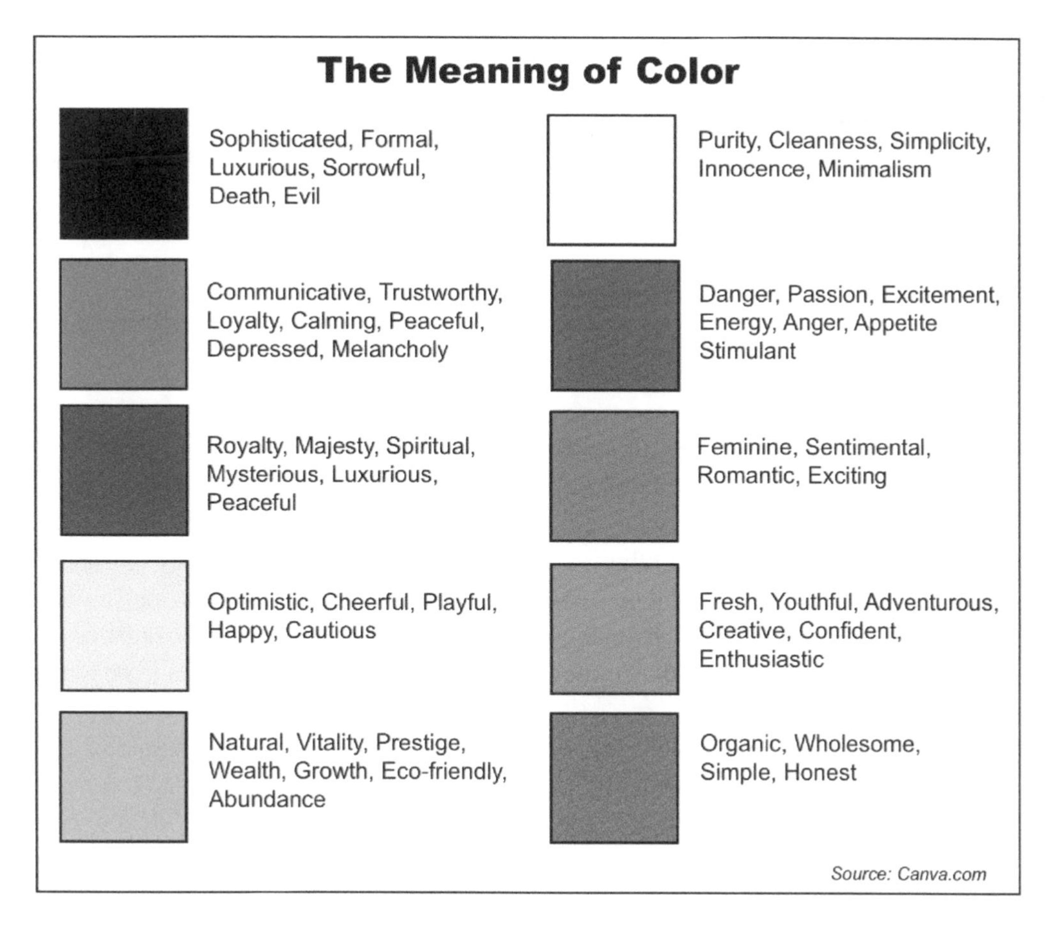


**FIGURE 1.10**

Again, the focus of this book is strategic *writing*; this is not a design textbook. However, a basic knowledge of design can improve your strategic communication, enhance your value as team member and increase your value to an employer.

# 1H Integrated Marketing Communications

Integrated marketing communications—more commonly known as IMC—is a valuable concept for strategic writers. The philosophy of IMC maintains that your target audiences receive many messages from your organization: ads, news stories triggered by news releases, speeches, tweets and other social media messages, and random exposures to products. IMC suggests that all those messages should be coordinated and "on-strategy." Otherwise, your target audiences will receive mixed messages from your organization and will become confused. To better understand IMC, let's look at each of its three words, beginning with *marketing*.

At its core, *marketing* means preparing a product that consumers want and helping them to acquire it. The "marketing mix," meaning everything that might persuade consumers, consists of product design, packaging, pricing, product demonstrations, ads and more. The marketing mix even includes the product's name. Years ago, marketing professors Philip Kotler and Jerome McCarthy defined the marketing mix with what they called the "Four P's" of marketing:

- Product (including name, design and packaging)
- Price
- Place (including putting the product where the consumer can buy it)
- Promotion (including tactics from advertising, public relations and sales and marketing)

Strategic writers sometimes believe marketing and the marketing mix mean only the fourth P, promotion. But marketing means much more.

Now for the word *communications*. The IMC philosophy says that each of the four P's communicates something to consumers. The name, design and package of a product send a message about quality to members of a target audience. So does price ("Wow, that's expensive; it must be luxurious") and place ("Hey, that's an upscale store; this must be a good product"). It's easy to see the communications aspect of the fourth P: Promotion directly involves the strategic writer, who will create ads, Facebook posts and other sales-related documents. In marketing, communication increasingly means two-way communication—almost a conversation between a company and a customer.

Finally, the word *integrated*. As you know, the IMC philosophy says that each of the four P's sends messages to consumers. The word *integrated* means that all those messages should work together. In other words, those messages should pursue a single strategy. The messages should not be haphazard and contradictory. For example, a luxurious product with a shockingly cheap price would send mixed messages to a target audience ("I thought that product was prestigious, but not at that price, I guess"). Or that luxurious product, now with a high price, would be out of place in a discount store ("I like that product, but it loses its prestige by being sold here").

Strategic writers should help create promotions with consistent messages. For example, that luxurious product shouldn't have an ad campaign that stresses one benefit (such as luxury), while a related public relations campaign stresses a different benefit (such as price) and a point-of-purchase campaign (in-store or online) stresses yet a third benefit, such as reliability. Who could blame the target audience for being confused about the product's image? Strategic writers should do more than simply ensure that all written messages are integrated; they should also help ensure that all the messages from the marketing mix—from the Four P's—are integrated.

Strategic writers often use a document called a strategic message planner (pages 171–191) to help develop the clear, consistent message that will be used to promote a particular product.

Recent studies show that the wide variety of social media outlets can pose a challenge to the integration of marketing messages. Slightly more than 80% of organizations say that their social media conversations are on-message—but that means that almost 20% of organizations remain challenged in this area.

There's more to IMC than just integration of messages. IMC practitioners:

- Focus on individual consumers. As much as possible, they develop products for individual consumers, and they create sales messages to target specific consumers' self-interests.
- Use databases to target individual consumers rather than huge audiences. These databases contain a wealth of information on individual consumers' wants, needs, preferences, media choices and buying habits.
- Send a well-focused message to each targeted consumer through a variety of communications, including mobile ads, news releases, texts, tweets and all other forms of marketing communication, including packaging and pricing.
- Use consumer-preferred media to send their marketing messages.
- Favor interactive media, constantly seeking informative conversations with individual consumers. New information goes into the databases mentioned earlier.

IMC is a logical extension of strategic writing. Strategic writing begins with the philosophy that a document's message should be on-strategy. IMC extends that philosophy to all related messages, ensuring that a variety of coordinated communications will send the same message to a target audience.

# 11 Ethics and Strategic Writing

Ethics are a combination of values and actions. In other words, being ethical means acting on our values (concepts to which we attach worth).

An ethics code, therefore, establishes guidelines for behavior. Ethics codes go beyond legal codes into the sometimes confusing world of right and wrong. Something legal, for example, isn't always ethical.

Knowing the right, values-driven course of action is often easy. But sometimes difficulty arises in *performing* that action. Sometimes, an unethical alternative can appear easier and less troublesome. For example, announcing and taking responsibility for a serious error—when that error is your own—can be difficult. In other cases, however, knowing the right, values-driven action seems impossible. In an ethics dilemma, our values seem to clash with one another, and every possible course of action seems to betray a value and cause unfair damage.

The origins of the word *ethics* suggest the challenges of behaving ethically. The Greek origin is *ethos* or *character*. But the earlier, Indo-European root of the word, according to the American Heritage Dictionary, is *s(w)e*—which means that related words include *secret, solitary, sullen, desolate, idiot* and even *suicide*. Even the history of the word *ethics* suggests the difficulty of ethical behavior.

## Rewards of Ethical Behavior

Ethical behavior is good for business. Although scholarly studies disagree about whether ethical behavior leads to financial success, they do agree on the opposite: Unethical behavior hurts profits and organizational success. Who would want to do business with crooks and liars? Poor ethics can interfere with the fulfillment of an organization's goals—especially long-term goals.

But there are other reasons for practicing good ethics. Ethical behavior is part of most of the world's religions. The philosophers Aristotle, Immanuel Kant and John Stuart Mill, who agreed on very little, all believed that we can never be truly happy unless we act on our values—in other words, unless we are ethical.

An emerging reward of ethical behavior involves a process called indirect reciprocity. Indirect reciprocity is a process in which one person or organization helps another with no likelihood or expectation of return and is rewarded by third parties who have observed and admired this behavior. In business terms, others seek economic relationships with your organization because they now view it as a reputable, reliable partner. The strategic value of indirect reciprocity was first noticed by evolutionary biologists and, in recent years, has been confirmed by studies in economics, psychology, anthropology and other human-behavior disciplines. New studies in psychology, in fact, demonstrate that individuals and organizations that practice indirect reciprocity earn more money, over time, than those that don't. Evidence from several disciplines now shows that helping others in need without the expectation of direct return is good for business.

## Ethics Codes

Ethics codes—written and unwritten—exist at several levels:

- International codes (such as the Caux Business Principles, www.cauxroundtable.org)
- Social or cultural codes (for example, the Ten Commandments)
- Professional codes, including the following: Public Relations Society of America (www.prsa.org); American Advertising Federation (www.aaf.org); American Marketing Association (www.marketingpower.com); International Association of Business Communicators (www.iabc.com)
- Organizational codes (such as the Credo of Johnson & Johnson, www.jnj.com)
- Personal codes (an individual's ethics code, written or otherwise)

In recent years, the ethics code of the Word of Mouth Marketing Association has gained increasing respect for its focus on establishing standards for the responsible use of social media within strategic communication.

Probably the ethics code that matters most to you is your personal code. As a writer, you know that writing down your thoughts stimulates precise thinking. Therefore, consider writing a personal ethics code, a document that specifies the core values that will guide your actions.

In writing your ethics code, you may wish to consider some of the great ethics principles of past millennia. In addition to focusing on principles established by important religious figures, university courses in ethics often emphasize these key philosophers and ideas:

- Both Aristotle and Confucius believed that virtue was a point somewhere between the extremes of excess and deficiency. For example, courage is the virtuous mean between cowardice and reckless bravery.
- Immanuel Kant believed that before committing ourselves to an action, we should ask ourselves if we would want to live in a world in which everyone did the same thing.

For example, could we live in a world in which *all people* broke their promises? Kant also believed that the end did not justify the means. In other words, he believed you couldn't justify a bad action that produced a good conclusion.

- Unlike Kant, John Stuart Mill believed that the end could justify the means. Mill believed that in an ethics dilemma, we should take the action that creates the greatest good for the greatest number of people. In other words, Mill believed the end *could* justify the means.
- John Rawls believed that justice involved fairness in the distribution of advantages and disadvantages. For example, he recommended that individuals who received unearned advantages such as good genes and birth into a stable upper-middle-class family should aid those who received unearned disadvantages in those areas.

These philosophies may help you act on the values specified in your ethics code. The best times to create and revise your code are when you are *not* facing an ethics crisis. In the depths of such a crisis, you'll need the clear, well-reasoned standards you established when you were free from doubts and fears. A crisis, of course, can prompt you to revise your ethics code.

## Ethics Challenges

As a strategic writer, you may face many challenges to ethical behavior:

- *Dilemmas*, in which every course of action will cause damage. Dilemmas occur when important values clash and it seems impossible to find one solution that honors all the involved values. A public relations agency, for example, might face a dilemma in wishing to part company with a disreputable client—but realizing that the separation would mean a loss of jobs within the agency.
- *Overwork*, which can lead you to inadvertently overlook important ethical considerations. A social media specialist facing a tight deadline, for example, might be tempted to use a photograph from an earlier event to represent something that just happened.
- *Legal/ethical confusion*, stemming from the dangerous belief that something legal is always ethical—and that something ethical is always legal. It's legal, for example, for an ad agency to work for two companies that compete with one another, but many professionals consider that practice to be unethical, particularly if the two companies are unaware of the situation.
- *Cross-cultural ethics*, in which important values from different cultures clash. The gift-giving norms of one business culture, for example, may clash with the gift-accepting norms of a different culture.
- *Short-term thinking*, which promotes a solution that postpones and increases pain and damage. Promising to meet an impossible deadline for the completion of a marketing

communications proposal, for example, might bring temporary rewards, but the cost of the inevitable failure probably would outweigh that short-term advantage.

- *Virtual organizations*, which consist of independent employees who temporarily unite to tackle a particular job. When the job is done, the organization dissolves. Can such organizations agree on ethical behavior? A similar challenge involves *freelancing*: If an established organization hires a temporary employee, will it remember to review its core values and ethics code with that individual? For example, a freelance social media assistant might, contrary to the company's value of honest and open communication, delete unfavorable comments regarding the company's Facebook posts.
- *New communications technologies*, which can lead to new ethics challenges. For example, programmatic advertising is a process that uses software and algorithms to automatically place ads in social media, other online media and older, traditional media. Programmatic advertising can be faster and cheaper than old-school human-to-human negotiations, but one result is ad placements that sometimes shock target audiences as well as advertisers. *Business Insider* magazine has documented an example of a cruise ship ad popping up in an online video of a sinking cruise ship and an example of a beer ad appearing in an article about underage drunk driving. Those situations almost surely offended some consumers and just as surely did not reflect the advertisers' values.

As you encounter these ethics challenges, remember that your mission is to build honorable, productive relationships that move your organization toward its goals. Ethical behavior isn't easy—but its rewards can be deep and lasting.

## Ethical Strategic Writing

As an ethical strategic writer, you probably will seek to build honorable, productive relationships between your organization and the groups with which it communicates. In building those ethical relationships, you may want to consider these values: honesty, completeness, timeliness and fair distribution.

### *Honest Documents*

- Context can affect honesty. For example, if you note that a production quota was met for the first time, you might also need to note that the quota had been reduced so that it could be met.
- Hard truths sometimes can hurt. Be diplomatic. Put yourself in the place of those affected by those truths.
- For legal and competitive reasons, documents cannot always contain every detail about every matter. But your documents should include accurate details in an honest context.

## Complete Documents

- As noted earlier, you can sometimes ethically withhold information. Justifications for withholding information include legal restrictions, individual rights to privacy and protection of competitive advantage. However, you should not withhold controversial information that a group has a right or a need to know. To paraphrase a famous politician, "If bad news will come out eventually—and it will—it should come out immediately."
- *Complete* is a relative term. In your documents, you should give groups the information they need and deserve. Most groups don't want to be buried in an avalanche of details. If there's any doubt about information a group needs, ask its members. Or use your best judgment and offer to provide more information upon request.

## Timely Documents

- Important information should be distributed quickly. This is especially true during crises for your organization or the groups with which it communicates.

## Fair Distribution of Documents

- The goal of distribution should be to reach every person who needs to see that message. Meeting this goal means using the communication channels preferred by recipients, which may not be the channels that you would prefer.

We hope you'll agree that ethical communication is not just the right thing to do; it's also a smart business decision. Strategic writing builds productive relationships. Good ethics are a strong foundation for productive relationships that will stand the test of time.

# 1J Diversity and Strategic Writing

Appreciation of diversity is essential for effective strategic communication. Many organizations aggressively seek employees from groups that sometimes are underrepresented in the workforce, including:

- Women
- Members of racial and ethnic minorities
- Physically and mentally challenged individuals
- Lesbian, gay, bisexual and transgender individuals
- Older workers

Organizations seek diversity for several reasons: Diversity is often a moral goal; it can be a regulatory requirement; it improves the organization's public image; it attracts the best employees; and it improves decision-making.

The authors of this book want to offer an additional benefit of diversity: Appreciation of diversity makes you a better strategic writer. Specifically, we recommend that strategic writers seek and study diversity in three areas:

1. Strategic communication staffs
2. Sources of information
3. Target audiences

Diversity in strategic communication staffs is more than a moral policy or compliance with regulations; it's a good business decision. Even within the same country, people from different cultural backgrounds have different values, concerns, hopes and communication traditions. For example, a word or image can be innocent to one person yet highly offensive to someone from a different culture. As international strategic communication becomes a daily reality, the danger of cross-cultural blunders increases. Having a diverse communication staff—and seeking knowledgeable partners in other nations—increases the likelihood of recognizing and preventing cross-cultural blunders.

Diverse staffs are more creative. Experts generally agree that creativity is fueled by the merger of different ideas—and a diverse staff, with its wide range of backgrounds, increases the diversity of the pool of ideas (see page 15). Innovations expert Jeffrey Baumgartner says that "diversity is the key to creativity."

Unfortunately, staff diversity continues to challenge strategic communicators: Fewer than half of public relations practitioners in a recent worldwide survey believed that their organizations reflected the diversity of the cultures within which they operated. Fortunately, three-fourths of those practitioners believed that their organizations were striving to increase the diversity of their communication staffs.

Diversity of sources also helps ensure successful strategic communication. In gathering research for the documents you write, draw upon diverse individuals—especially those you will quote or cite by name. Sources whose backgrounds offer different views of a particular issue might provide valuable new insights that help you achieve your strategic goal. In many organizations, it's easy to rely on a steady stream of white Anglo-Saxon males in their 40s and 50s. Although these men aren't necessarily bad or even similar sources, a more diverse group of sources might offer more perspectives as well as more appeal to target audiences.

The diversity of target audiences can vary widely. Although strategic writers should carefully study the values and concerns that unite members of a target audience, they also should be aware of differences within the target audience. To address a large group, strategic writers often must focus on the values of the majority. Whenever possible, however, strategic writers must know and avoid things that might alienate different minority groups or individuals within the target audience. Strategic writers should avoid stereotyping members of the target audience.

Strategic writers can show appreciation for diversity by following these five guidelines:

1. Use *they* instead of *he* or *she* to describe a generic individual. This used to be considered bad grammar, but it makes so much sense that Associated Press style now recommends it. It's OK to write something like *When an employee offers their opinion . . . .* If you're old-school and that sounds incorrect, you can still avoid the *he/she* challenge by making the relevant words plural: *When employees offer their opinions . . . .*
2. Avoid words that unnecessarily describe particular relationships: *your wife, your husband, your boyfriend, your girlfriend, your parents* and *your children*. Some readers may no longer have parents; some may not have children. Some may be unmarried.
3. Know the dates of major religious holidays and events. When is Rosh Hashanah? When is Ramadan? When is Easter?
4. Don't describe individuals by race, ethnicity, religion, age, sexual orientation or physical or mental characteristics unless the information is relevant to your document's purpose. If an individual must be so described, consider applying the same exactness of description to every other individual mentioned in the document.

5. If you are responsible for a document's design, apply your quest for inclusiveness to photographs and other visual representations of individuals. Even if you're *not* in charge of the design, don't hesitate to point out lapses of diversity.

The array of terms for ethnic and racial groups can present challenges for strategic writers. A landmark U.S. Bureau of Labor Statistics survey of almost 60,000 households revealed these preferences:

- Blacks prefer *black* (44%) to *African American* (28%) and *Afro-American* (12%).
- Hispanics prefer *Hispanic* (58%) to *of Spanish origin* (12%) and *Latino/Latina* (12%).
- American Indians prefer *American Indian* (50%) to *Native American* (37%).
- Multiracial individuals prefer *multiracial* (28%) to *mixed-race* (16%).

The best policy may be to ask the individuals you're writing about how they prefer to be identified—if such identification is necessary.

Diversity is an asset for your organization and its communications. Be aware of its value in your quest to be a successful strategic writer.

# 1K Persuasion and Strategic Writing

Persuasion is a controversial concept. People sometimes see persuasion as a win-lose game: One side wins, and the other side loses. However, persuasion in strategic writing works best when it promotes a win-win scenario. Effective strategic writing seeks benefits for all sides in a relationship.

How can strategic writers create persuasive win-win scenarios? One way is to understand the target audience. Dean Rusk, a former U.S. secretary of state, once said, "The best way to persuade anyone of anything is to listen." When strategic writers listen to the hopes, fears, concerns and desires of their target audiences, they are better prepared to create strategic messages that satisfy both the organization and the target audience. Strategic writers who listen can help shape persuasive messages that unify rather than divide.

As you listen to members of a target audience, seek an answer to the all-important question of WIIFM: What's in it for me? In other words, how will members of the target audience benefit from the information in your document? What's in it for them? If you're writing an advertisement, how will consumers gain from purchasing your product? If you're writing a news release, why is the news important to journalists and their audiences? If you're preparing a Facebook status update, why should people care? Imagine that every target audience is ready to greet your message with two shouted responses: "So what? What's in it for me?" If your message can answer these questions and present benefits to the target audience, you can probably reverse the process and get *them* to listen to *you*.

In some situations, listening becomes dialogue—and dialogue becomes negotiation and persuasion. You do this frequently when you listen to a friend, consider their concerns and attempt to move them toward an action that will benefit you both. Communication scholars George Cheney and Phillip Tompkins have developed four principles that they believe should guide persuasive negotiations:

1. **Empathy**. You should truly listen, motivated by a desire to find a solution that's best for everyone.
2. **Guardedness**. Just because you're willing to listen doesn't mean you have to agree and change your opinions.

3. **Accessibility**. However, you should be willing to consider changing your opinion. Consider that you might be wrong.
4. **Nonviolence**. Threats have no place in ethical persuasion.

By following these four principles, you can support your organization and still keep an open mind as you search for a win-win solution.

Just as listening is key to successful persuasion, so is your character as a persuader. Personal credibility is one of the most powerful tools of persuasion. Almost 2,500 years ago, Aristotle wrote that there are three approaches to persuasion: *logos* (an appeal to the target audience's intellect), *pathos* (an appeal to the target audience's emotions) and *ethos* (an appeal based on the speaker's character). Communication scholars today still agree with Aristotle's analysis. Of those three approaches, Aristotle wrote that *ethos* was usually the most powerful. (We've all heard that "virtue is its own reward," but in negotiations, virtue provides an additional reward: persuasive power.) As you learn more about a target audience, you should consider what combination of *logos*, *pathos* and *ethos* would be most persuasive.

## Monroe's Motivated Sequence

Almost a century ago, communication scholar Alan Monroe developed a blueprint for persuasive messages. Today, many strategic writers use "Monroe's Motivated Sequence," which consists of five parts:

1. **Attention**. Grab the target audience's attention. Chances are, the target is overwhelmed by messages. Cut through the clutter. Get noticed.
2. **Need**. Describe an important problem that the target audience faces—a problem that needs a solution. (You'll discover this need/problem by listening.)
3. **Satisfaction**. Offer a solution that benefits both you and the target audience.
4. **Visualization**. Illustrate the consequences of success—and the consequences of failure. (This often can be the most powerful, motivating part of Monroe's Motivating Sequence.)
5. **Action**. Tell the target audience what it can do to enact the solution.

Monroe's Motivated Sequence can work in several documents described in this book: speeches, memos, ads, sales messages, proposals and many more.

## Syllogisms and Enthymemes

Another powerful tool is Aristotle's enthymeme, which is built upon a syllogism. We often reach conclusions via syllogisms. A syllogism consists of a major premise (a generally accepted truth or value), a minor premise (a specific claim) and a consequent conclusion.

For example:

| | |
|---|---|
| **Major premise (an audience belief)**: | I need a phone that won't crack when I drop it. |
| **Minor premise (your message)**: | Brand X phones are the toughest in the market, built to withstand 25-foot drops. |
| **Conclusion (an audience decision)**: | I need a Brand X phone. |

In an effective enthymeme, strategic communicators understand the target audience so well that they simply deliver the minor premise (message)—no major premise or conclusion. They know that, with a little luck, the audience will link their message to the relevant major premise—and will then draw its own conclusion. With a successful enthymeme, the audience persuades itself; the audience creates its own call to action.

Persuasion is unavoidable in strategic writing. Ethical persuasion, based on listening and seeking win-win relationships, can create enduring, successful relationships.

# 1L The Law and Strategic Writing

In the United States, courts have ruled that strategic writing is similar to speech—and freedom of speech is protected by the First Amendment of the U.S. Constitution:

> Congress shall make no law respecting an establishment of religion, or prohibiting the free exercise thereof; or abridging the freedom of speech, or of the press; or the right of the people peaceably to assemble, and to petition the government for a redress of grievances.

Although strategic writing enjoys some constitutional protection, that doesn't mean that there are no restrictions. For example, you can't legally create, publish and distribute obscene material. You can't legally threaten national security. Even free expression has its limits.

Legal restrictions on strategic writing depend on the purpose of the writing. U.S. courts have ruled that free expression falls into two general categories: political speech and commercial speech. When your organization expresses its opinion on a social issue—such as support for literacy programs—that communication is political speech. Political speech faces few legal restrictions because the courts have ruled that the free exchange of ideas is essential to democracy. However, when your organization communicates in pursuit of a financial goal—such as an advertisement designed to sell your product—that ad is commercial speech. In general, commercial speech faces more legal restrictions than political speech does.

Communication laws constantly evolve, and strategic writers need to keep up with those changes. Traditionally, writers should keep updated in six areas of communication law: libel, invasion of privacy, deceptive advertising, electioneering, copyright and financial disclosure. These six areas don't divide neatly between political speech and commercial speech. Advertising, for example, usually would be commercial speech. However, an ad that supports local literacy programs might be considered political speech. That is why it is important to know the law or contact a lawyer who does.

Libel is "injury to reputation." If a strategic document includes an untrue claim that exposes an individual to public scorn, hatred or ridicule, then that message may be libelous—and, therefore, subject to legal action. For a message to be libelous, it must have these qualities:

- *Defamation*. The message must expose an individual to public scorn, hatred or ridicule.
- *Publication*. The message must be published. However, courts have defined publication very broadly. Sending a text to another person can constitute publication.
- *Identification*. The message must identify the defamed individual. However, identification need not be by name. If a description of the individual is so complete that a name is unnecessary, identification has occurred.
- *Negligence*. The message must be inaccurate, and there must be little excuse for that inaccuracy. However, public figures such as politicians have a higher burden of proof for negligence. The U.S. Supreme Court has ruled that public figures must prove that the writer acted with *actual malice*, which is defined as "knowing falsehood or reckless disregard for the truth."
- *Damage*. The message must have damaged the identified individual. Damage can be as specific as financial losses or as vague as loss of reputation.

As databases provide more and more access to information about target audiences (see Integrated Marketing Communications, pages 42–43), strategic writers must understand the legal limits of acquiring and communicating personal information—both in the United States and elsewhere. This is especially true for organizations that do business in the European Union. The General Data Protection Regulation (GDPR) is the toughest privacy law in the world. GDPR levies harsh fines against those who violate the regulations even if they are not located in the EU. (Google was fined $57 million in 2019 for improper disclosure to consumers.) The regulation was created with the premise that people have a right to control their personal information.

The U.S. Constitution doesn't specifically guarantee a citizen's right to privacy. However, communication law does protect some areas of privacy. Illegal invasion of an individual's right to privacy usually comes in one of four forms: *intrusion* (a physical violation of one's privacy), *false light* (portraying an individual in an improper and/or unfair context), *publication of private facts* (disclosure of information that, in a legal sense, is private), and *appropriation* (the use of someone's name, voice, likeness or other defining characteristic without permission).

New legislation affecting social media can touch on privacy issues. For example, the Telephone Consumer Protection Act mandates that companies using mobile-messaging ads must secure recipients' written permission before sending sales-related texts, must offer an easy "opt-out" procedure and must remind recipients that "standard message and data rates may apply."

In the United States, deceptive advertising doesn't necessarily mean untrue advertising. In this country, advertisers can legally stretch the truth—within limits. For example, a new toothpaste may seem to promise teeth so white that passersby will gasp in admiration and write you love poems. Do consumers really believe that? Probably not: The exaggerations are just a humorous way to make a point. Acceptable exaggeration in advertising is called *puffery*.

But what happens if that same toothpaste ad falsely claims that the product prevents sore throats? If the ad presents that falsehood in a way that could fool a reasonable consumer, it becomes deceptive advertising. The U.S. Federal Trade Commission determines when puffery deviates into deceptive advertising, and it has a range of punishments it can apply, including forcing the company to run corrective advertising that retracts false claims. To separate puffery from deceptive advertising, the FTC uses the "reasonable consumer standard." That standard says that if an ad could deceive a reasonable consumer who expects exaggeration in advertising, the ad becomes deceptive and, therefore, illegal.

Laws regarding puffery and deceptive advertising differ from country to country—as do laws regarding libel, invasion of privacy, elections and other key legal areas outlined in this chapter. Globally, there is increased scrutiny of marketing and advertising directed toward children. In fact, in Norway and Quebec, advertising to children under the age of 12 is illegal. Writers who help create international strategic communication campaigns should be aware of those differences.

Laws governing what the Federal Election Commission refers to as "electioneering communication" constantly change. Not all electioneering communication is the same, as the FEC has defined several different categories. However, all have specific disclosure requirements. Strategic writers can find the latest regulations on the FEC website.

Companies that sell stock—publicly owned companies—are subject to disclosure law. Disclosure law governs how and when companies should communicate about matters that affect or could affect their stock prices. Its purpose is to ensure that every investor has a fair and equal chance in the stock market. Publicly owned companies must provide quarterly financial reports (Form 10Q) and annual financial reports (Form 10K) to the U.S. Securities and Exchange Commission. They also must provide annual reports and annual meetings for their stockholders. Publicly owned companies must notify the SEC of unusual events that could affect stock prices (Form 8K). Those companies also must notify stockholders about such news, usually through news releases sent to relevant news media. The Sarbanes-Oxley Act of 2002 mandates that leaders of publicly held companies are personally responsible for the truthfulness of their companies' financial statements. All disclosure-related communication must comply with SEC Rule 10b-5, which prohibits misleading statements in all such communications.

Copyright law protects intellectual property. Federal law defines *intellectual property* as "original works of authorship that are fixed in a tangible form of expression." For strategic writers, copyright law regulates the use of someone else's creative work. To use such work, we must gain the creator's written permission. This often involves paying a fee established by the work's creator. Exceptions do exist: The "fair use" doctrine allows students, teachers, reporters, reviewers and others to use copyrighted works to inform others. However, fair use generally does not protect those who borrow intellectual property for commercial purposes. Even if fair use does protect such borrowing, strategic writers should always credit, in writing, the creator of the work. Although copyrighted material often is marked by the symbol ©, a work has legal copyright protection from the moment its creator establishes it in a fixed, tangible form. You also should know that work completed for an employer generally is "work for hire" and legally belongs to the employer.

It is important to remember that laws pertaining to old media also apply to newer media as well. When someone tells you "everything on the internet is free," that is not true. Rules covering copyright, trademarks, libel and financial disclosure remain in force. For example, as early as 2013 the U.S. Securities and Exchange Commission issued guidelines covering the use of social media, ruling that companies could use Facebook and Twitter for financial disclosure as long as investors knew in advance of such use.

When it comes to communication law, get advice from experts. Remember this adage: "A person who serves as their own lawyer has a fool for a client."

# 1M Jobs in Strategic Writing

Don't let the brevity of this segment alarm you: Jobs in strategic writing *do* exist. As long as organizations communicate, strategic writers will have jobs. Good strategic writers might have choices among good jobs. And really good strategic writers might discover an irony in their careers: Their successes in writing might pull them away from writing. A great ad copywriter might become an agency's creative director—more memos, proposals and business correspondence, certainly, but fewer ads. A talented writer of stories for an online magazine might soon become the editor of the magazine—more money, but less writing of informative, entertaining features. These talented individuals will groom the next generation of strategic writers, searching for writers as good as themselves. Good writers recognize and value good writing.

So where do you start? Prove that you're a good strategic writer. Seek internships and volunteer opportunities that will allow you to write. Do real writing for real clients. Enjoy your successes and learn from your failures. Pay attention to your professors. Collect your best work in an online portfolio that will impress potential employers.

Every organization needs strategic writing. Some do it in-house, others hire agencies and many combine both approaches. One good way to find a satisfying job in strategic writing is to pick a geographic area that interests you—say, Tierra del Fuego (we hope you *hablas español*). Now find out what organizations have offices in Tierra del Fuego. Begin your job search by studying and applying to those organizations.

Another job-search strategy involves combining your talent in strategic writing with a passion in your life. One of our favorite graduates double-majored in public relations and art history. She's now marketing director for an art museum.

The U.S. Bureau of Labor Statistics estimates that approximately 1 million people work in various aspects of strategic communications in the United States. In its most recent *Occupational Outlook Handbook*, the BLS estimates that there are approximately 270,000 entry-level jobs in public relations and almost 290,000 management-level jobs in advertising and PR—and the bureau expects those numbers to steadily rise through 2028. Successful job candidates, according to the BLS, will have internship experience, a portfolio of real work for real clients and strong writing skills.

Jobs in strategic writing tend to cluster in five categories: corporations, agencies, government agencies, nonprofit organizations and trade associations and independent consultancies. Let's quickly look at each.

## Corporations

Corporations are for-profit businesses that can be as large as General Motors or as small as a local dry cleaner. This book focuses on four areas of strategic writing: public relations, advertising, business communications and sales and marketing. Corporations, especially large ones, hire strategic writers in all four areas. Because corporations are for-profit businesses, they have relationships with a wide variety of groups: customers, employees, government regulators, the news media, stockholders and many more. Developing and maintaining these relationships requires strategic writing. Of these five employment categories, corporations tend to pay the highest starting salaries for entry-level strategic writers.

## Agencies

Agencies supply advice and strategic communications for other organizations. Some agencies are international, with offices throughout the world. Others operate from spare bedrooms in suburban houses. Three broad categories exist: advertising agencies, public relations agencies and full-service agencies that combine advertising, public relations and other sales and marketing functions. Because agencies are businesses, they also rely on business communications: reports, memos, business correspondence and more. Of the five employment categories, agencies tend to have the second-highest starting salaries for strategic writers.

## Government Agencies

Government agencies exist at the international, national, state and local levels. They can be as big and well-known as the U.S. Securities and Exchange Commission or as small as your local school district. Strategic writing can be a diplomatically sensitive subject for government agencies. If they openly engage in public relations, advertising and marketing to promote themselves or elected politicians, members of the voting public may cry, "Propaganda! Waste of taxpayers' money!" Yet some government programs and projects must be promoted, such as the U.S. Department of Agriculture's food pyramid. Strategic writers for government agencies must constantly be aware of the gray area between legitimate strategic communication and unacceptable promotion. Of the five employment categories, government agencies tend to have the third-highest starting salaries for strategic writers.

## Nonprofit Organizations and Trade Associations

Nonprofit organizations provide services without the expectation of earning a profit, though they do require money, of course, to fund operations. They can be as large as the World Wildlife Fund or as small as the local community college. Trade associations resemble nonprofit organizations in that they offer services without the primary motive of profit. Trade associations include such groups as the National Association of Home Builders. Like corporations, nonprofit organizations and trade associations traditionally have strategic writing positions in public relations, advertising, business communication and sales and marketing. And because these organizations generally have smaller communications staffs than do corporations, they offer great opportunities for writers who don't want to specialize. A strategic writer for a small nonprofit organization may work on a news release, a mobile ad, a proposal, a fundraising letter and social media posts all on the same day. Of the five employment categories, nonprofits tend to have the fourth-highest starting salaries for strategic writers.

## Independent Consultancies

An independent consultant is a freelancer. A freelancer may often be part of a "virtual organization," a business group that forms for one project and then disbands. Independent consultants generally specialize in one of the professions discussed in this book—social media marketing, for example. However, versatility can mean more clients and more profits. In addition to strategic writing, independent consultants carry the burden of finding clients, answering the phone, filing, making coffee and finding time for a life. Of the five employment categories, we rank independent consultancies last in starting salaries—and, in a sense, that's unfair. Successful consultants often earn more than the average corporate strategic writer. However, very few strategic writers begin as independent consultants. Instead, they work for other organizations, learn the ropes, earn a reputation—and then take a deep breath and go out on their own. Consultants usually are experienced professionals. The concept of starting salary takes on a different meaning for consultants.

Whichever job area appeals to you most, you can increase your odds of employment by marketing yourself: Get experience through internships, part-time jobs and real work for real clients. Join and participate in established networking sites such as LinkedIn. Blog about a topic you enjoy and know a lot about—and use Twitter, Facebook and other social media to drive readers to your blog. Build a great portfolio and a great résumé.

We believe in the value of strategic writing. We believe in the value of *jobs* in strategic writing. So work hard. Do research. Stay on message. Get experience in both old and new media. Build a great portfolio. Somewhere out there, a good job in strategic writing awaits you.

SECTION

2

# Strategic Writing in Public Relations

## OBJECTIVES

**In Section 2: Strategic Writing in Public Relations, you will learn to write these documents:**

- Microblogs and status updates
- Blogs
- Podcasts
- Website pages
- Traditional news releases
- Feature news releases
- Promotional news releases
- Media advisories
- Pitches
- Video news release scripts
- Media kit backgrounders
- Media kit fact sheets
- Photo opportunity advisories
- Newsletter and magazine stories
- Annual reports
- Speeches

# 2A Introduction to Public Relations

Public relations often gets confused with publicity. Public relations certainly includes publicity—but it includes much more. A standard definition of *public relations* shows how broad the profession can be: Public relations is the values-driven management of relationships with publics that are essential to an organization's success.

As we noted in Section 1, well-run organizations often have brief statements that specify their most important values. One challenge for public relations practitioners is to build relationships that honor not only those values but also the values of the other group in a relationship, such as the values of journalists, stockholders or employees.

Well-run organizations also have goals consistent with their values. To reach those goals, the organizations often need resources that they don't control. For example, to reach its goals a major corporation needs resources held by employees (the willingness to work hard), stockholders (the willingness to buy stock and to vote for the managers in annual meetings), the news media (the willingness to cover the organization fairly) and many other resources held by other groups. Public relations practitioners strive to develop positive, productive relationships with publics that control essential resources. Those relationships are productive when the organization receives the needed resources.

The term *public* has a specific meaning in public relations. A public is any group whose members have a common interest or common values in a particular situation. A public can be as official as the members of a state legislature or as unofficial as the residents of a neighborhood where your organization wants to build homes for Habitat for Humanity.

When a public can be affected by the actions of an organization (or vice versa), that public is said to have a *stake* in that organization. Such publics often are called *stakeholders*.

Many of us think of public relations as a process of getting different groups to like our organization—of getting our organization's message to people and making them agree with it. Just like publicity, that can be part of public relations. But, again, public relations is bigger than that. It's about building productive relationships with resource-holders. And that's definitely easier when those publics like you and agree with you.

The actual practice of public relations follows a process that we also find in advertising, business communication and sales and marketing. That process consists of four stages: research, planning, communication and evaluation. As straightforward as that seems, sometimes the process needs to move backward. For example, as we create a plan based on our research, we may discover that we need more research before we can finish planning. Or as we communicate in accordance with our plan, the situation may change, forcing us to go back to the planning stage. Evaluation is a form of research that can lead to more planning. But in general, we conduct research and we plan before we communicate.

According to public relations scholars Betteke van Ruler and Dejan Verčič, PR professionals have four basic roles:

- Counseling their organization's leaders on the importance of productive, win-win relationships with publics that hold key resources
- Creating research-based plans to build positive relationships with those publics
- Carrying out those plans
- Coaching all members of their organization to behave in ways that build good relationships with key publics

Note the importance of planning within those four roles. At the heart of a good PR plan are strategic messages delivered to particular publics. To develop and deliver strategic messages, PR professionals often use a planning grid that seeks information and ideas in 10 distinct areas:

**Plan Goal:** ______________________________

| Public | Resource | Value(s) | Relationship | Message | Media | Timetable | Manager | Budget |
|---|---|---|---|---|---|---|---|---|
| | | | | | | | | |
| | | | | | | | | |
| | | | | | | | | |

1. The **goal** of the plan specifies what general relationship/resource outcome you hope to achieve.
2. A **public**, as you know, is any group whose members have a common interest or common values in a particular situation.
3. A **resource** is something a public has that you want. For example, if your organization wants local news media to cover a special event that you're hosting, those news media have the resource of coverage.
4. **Values** specify the core ideas or beliefs that unite members of the particular public.
5. **Relationship** concisely describes the nature of the contacts, if any, between your organization and the particular public. For example, is it an ongoing, positive, productive relationship? Or is it, unfortunately, sporadic, negative and distrustful?

6. The **message** is key: A good strategic message addresses the values of the public in a way that will lead its members to deliver the desired resource that advance you toward your goal. A good message shows the public how it can honor its own values by sharing its resources with your organization. A good message takes into account the nature of the relationship with the public.
7. **Media** specify the channels you'll use to send the message. Of course, those should be the media your target public prefers.
8. **Timetable** specifies when you'll send the message.
9. **Manager** specifies who on your team has the responsibility to ensure the creation and delivery of the message.
10. **Budget** specifies how much money you can spend on the creation and delivery of the message.

Again, the goal of public relations is to build and maintain resource-delivering relationships. However, van Ruler and Verčič emphasize that such relationships can't happen unless the communities that surround an organization view it as a good citizen. Successful organizations, therefore, tend to practice a philosophy known as corporate social responsibility (whether the organization is a corporation or not). Corporate social responsibility has many definitions, but most involve the idea that an organization should sustain and improve the societies within which it operates.

A recent survey of senior public relations professionals throughout the world found that good writing remains the most important job skill within public relations, followed by strategic planning abilities, speaking abilities and data analysis expertise.

# 2B Social Media in Public Relations

## Purpose, Audience and Media

Hootsuite, an international social media management company, reports that we now spend 2.5 hours every day using social media. Social media now reach 49% of the world's population. Strategic communicators need to ensure that their organizations are involved in those conversations.

The audiences for social media within public relations are vast and diverse. Social media usage is increasing among all key publics: customers, employees, journalists, advocacy groups, investors, government workers and even older adults. For example, the Pew Research Center reports that a majority of U.S. citizens now monitor current events via social media.

The most-used social media platforms in the United States vary from public to public and, probably, from month to month. Adobe, however, names six platforms that will be essential for public relations professionals in the early 2020s: Instagram, YouTube, Facebook, Twitter, Pinterest and Snapchat.

As social media grow in importance in public relations, four important trends within these ever-changing platforms show signs of enduring: the relevance of social media influencers, the importance of monitoring social media conversations, the use of algorithms and the importance of video.

## Social Media Influencers

Social media influencers are opinion leaders. Members of publics often turn to particular individuals and organizations for advice about purchases, social issues and much more. In fact, a recent Twitter survey found that 49% of Twitter users consulted specific influencers (who were not friends and family) within that platform before making significant purchases. Given the importance of these opinion leaders and the new realities of social media algorithms (see below), Michael Stelzner, founder of the website Social Media Examiner, counsels PR professionals to reduce their obsession with measurements such as likes and retweets and turn more attention to building relationships with carefully chosen social media influencers. Companies such as BuzzSumo can help you identify primary influencers for key publics—but don't hesitate to directly ask members of those publics to whom they turn, within social media, for advice.

## Social Media Monitoring

Monitoring social media conversations may be just as important as joining them. Social media monitoring generally has two missions: (1) carefully listening for any mentions of your organization, especially those that demand a response; and (2) listening for any issues-oriented conversations that your organization might join in ways that could serve important publics such as journalists, stockholders or potential employees. Several companies, including Talkwalker and Sprinklr, specialize in offering social media listening and analysis services.

## Social Media Algorithms

Let's keep this simple. Algorithms within social media platforms matter because they're top-secret mathematical formulas developed by each platform to determine which messages people will see. For example, you probably realized long ago that you don't see every status update from everyone you've friended on Facebook. Instead, Facebook algorithms sort out which posts—in Facebook's opinion—will be of most interest to you. Twitter's "You might like" and "In case you missed it" clusters are driven by algorithms. For PR professionals, this means that not every member of your target publics—even those who follow you in social media—will see your messages, even if they're online when you post. Best practices for coping with social media algorithms include encouraging one-to-one conversations via such things as direct messages and comments and staying current regarding what kinds of posts algorithms tend to favor—which leads to our third trend: videos.

## Videos

Social media algorithms currently favor posts with videos, a trend that, again, shows signs of enduring. Live videos fare even better with current algorithms. Options such as Facebook Live and Periscope, thus, should remain primary considerations within public relations plans. Sprout Social, a social media management company, reports that selection algorithms increasingly favor video content, a trend that will continue into the foreseeable future. Demographic studies consistently show that users in their teens and 20s generally seek entertainment videos. Older groups seek how-to and informational videos.

The rewards for keeping up to date with social media practices are significant: A recent survey of international practitioners ranked social media skills second only to the creation of strategic content among the drivers of future growth for the profession of public relations.

# 2c Microblogs and Status Updates

## Purpose, Audience and Media

Microblogs, such as tweets and Facebook status updates, help organizations and individuals maintain ongoing, productive conversations with key publics. The "Twitter Basics" section of Twitter.com declares:

> Twitter is where people go to find out what's happening in the world right now. Whether you're interested in music, sports, politics, news, celebrities, or everyday moments, come to Twitter to see what people are talking about and join the conversation.

Instagram's mission is "to share the world's moments." Clearly, microblogs are for communicating and building relationships. They also can serve as aggregators, providing links to information of interest to targeted publics—especially important in content marketing (pages 245–246).

Traditional publics for microblogs and status updates include any groups or individuals that possess desirable resources: customers, employees, stockholders, voters and more. Publics for microblogs and status updates, however, generally are self-selecting. Most organizations, therefore, use many forms of media, including their websites, to encourage people to follow them on Twitter, Facebook and other social media platforms. Demographic studies show that Twitter and Facebook usage is greatest among individuals aged 18–45. Older publics use social media less, although their usage increases every year.

Microblogs and status updates exist only digitally. They appear as concise messages on computer or mobile-device screens.

**KEY TO SUCCESS**

Business-oriented microblogs and status updates are brief, highly focused and often multimedia. They immediately emphasize something of interest to the reader.

## Format/Design

Compared with Twitter, social media such as Facebook and LinkedIn include several sections for information: Photos, Videos, Events, About Us and more. However, the formats and capabilities of social media vary and change rapidly.

Profile photos or icons usually accompany and identify Facebook status updates and tweets. Organizations generally use their official logos. Starbucks, for example, uses its familiar green mermaid. Rock groups such as the Foo Fighters often use artwork from a recent album.

In addition to profile images and brief written messages, status updates and tweets often include an attention-grabbing photo, video, GIF or link image. Survey research confirms that posts with visual elements gain more attention than do simple text posts.

Because of the increasing popularity of microblogs and similar media, several services have emerged that permit users to coordinate posts to many social networks. For example, companies such as Hootsuite allow you to manage posts to Facebook, Twitter, Instagram and other popular social media platforms.

## Content and Organization

Although business-related microblog posts, such as Twitter tweets, and status updates, such as those on a Facebook page, are different, they do have much in common:

1. They're short. The maximum length for Twitter tweets is 280 characters and spaces. Facebook status updates generally shouldn't exceed four sentences.
2. They quickly focus on reader interests. (Remember that your target publics will focus on WIIFM: What's in it for me?)
3. They encourage responses, either through comments, replies and/or links to websites.
4. They often are responses to messages initiated by others. Consider commenting on and/or sharing favorable messages posted by others. In Facebook, you can comment or simply click "Like" or one of the related emojis to respond to someone else's post. In Twitter, you can retweet favorable comments posted by others. Within no more than 24 hours, be sure to respond to questions and negative comments.
5. They're conversational, but they use strong verbs and precise nouns to achieve brevity and power.
6. They use casual grammar, including sentence fragments. However, such informality must clearly be intentional, not a careless error. When appropriate, they show a sense of humor. For a good example, check out Arby's multimedia Twitter presence.
7. They use personal pronouns: *you, your, we* and *our.*
8. They avoid jargon and technical language that could puzzle some readers.

9. They use capitalized words for emphasis, but they DON'T overdo it.
10. They can include links to news releases and other documents and sites, including digital newsrooms (pages 132–134) and YouTube videos.
11. Finally, they avoid using vulgar or obscene language, images or links.

Twitter in particular includes a variety of ways to encourage conversation:

a. *Mentions*: Twitter users have usernames. By placing the symbol @ before an individual's or organization's username—for example, @username—you can ensure that the tweet will go either to the user's Twitter page (if they follow your account) or a folder that contains such messages. Such messages are not private; they're available to anyone who follows both the sender and the recipient or simply visits the sender's profile.
b. *Replies*: To reply to a tweet, click or tap on the Reply icon (a left-pointing arrow) beneath the original tweet. Such messages also are public.
c. *Direct messages*: You can send private tweets—direct messages—to any user who follows you either by clicking or tapping on the Envelope icon. Direct messages also can be used for private group conversations. Organizations can adjust their Twitter accounts to receive private, direct messages from users whom they don't follow.
d. *Retweets*: You can share—retweet—others' messages by clicking or tapping on the Retweet icon (two curved arrows) at the bottom of a tweet. You can then add a comment or simply share the original tweet; adding a comment to a retweet is known as quote tweeting.
e. *Hashtags*: You can insert hashtags into your tweets to help categorize them and make them easily findable in keyword searches. Just add a hash mark (#) to a keyword—for example, *#publicrelations* or *#BTS*. Hashtags often appear at the end of tweets—and some are more for editorial comment (for example, *#eyesrolling*) than category purposes. Erin E. Templeton, a frequent poster in the ProfHacker blog, recommends concise hashtags and notes that groups that really do want to use hashtags to share messages should create unique hashtags that everyone in the group knows and will use. Trending topics within Twitter often are hashtags.
f. *Shortened website addresses*: Website addresses, also known as URLs, can consume a big percentage of your 280 characters and spaces. If you tweet from Twitter.com, that site often will shorten the URL for you. Google's URL Shortener app can do the same.

Microblogs are ideal media for delivering engaging, informative and entertaining content to important publics. Platforms such as Facebook, Instagram and Twitter can be important parts of content marketing (pages 245–246).

## MICROBLOG TIPS

1. **Go to the experts**: Microblog and social network services often have Help sites for users. For example, the Facebook Newsroom offers tips and updates, and Twitter has a similar Twitter Help Center.
2. **Imitate the best**: Study how other organizations use microblogs and status updates. What works well—and what doesn't?
3. **Plan**: Build a social media calendar, and coordinate your posts with a social media dashboard (pages 247–251).
4. **Be timely**: Post during times when your research shows that your target publics are most likely to be online. Organizations such as Sprout Social often publish studies about the best times of day to post for different target audiences.
5. **Build a team**: Designate managers for organizational microblogging and status update functions. Your organization should have clear gatekeepers for such messages.
6. **Expand**: Facebook, Twitter and Instagram are current leaders, but be present in LinkedIn, Snapchat, Pinterest or any social media preferred by target publics.
7. **Monitor**: Continually scan social media for references to your organization. Have a policy for addressing criticism and removing comments from your own sites. And have a thick skin; it can be rough out there.

# Twitter Tweet

Credit: green_01/shutterstock.com; michaeljung/shutterstock.com

## Twitter Link to News Release

Credit: green_01/shutterstock.com; oneinchpunch/shutterstock.com

## Twitter Retweet

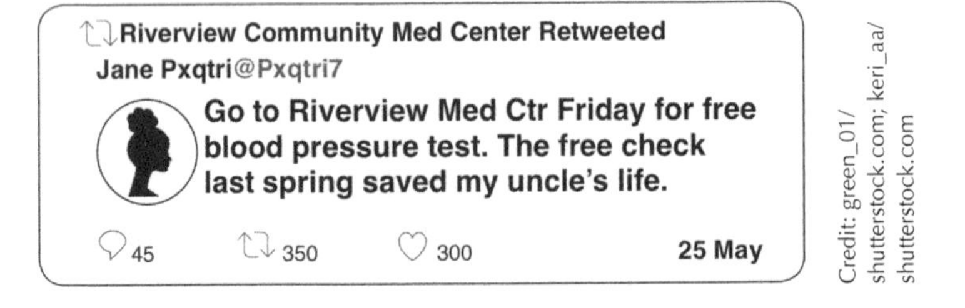


Credit: green_01/shutterstock.com; keri_aa/shutterstock.com

A simple retweet does not include your organization's icon.

## Twitter Retweet With Comment

Credit: green_01/shutterstock.com; michaeljung/shutterstock.com

Adding a comment to a retweet adds your organization's icon and diminishes the icon of the original tweeter.

## Facebook Status Update

CREDIT: MICHAELJUNG/SHUTTERSTOCK.COM

# 2D Blogs

## Purpose, Audience and Media

Blog is short for weblog, an online personal essay. Blogs are informal, informative and often entertaining: They seek to build relationships with regular readers. They sometimes resemble opinion columns in news sources: Both provide individual responses to current news stories as well as to other issues—everything from favorite recipes to terrible first dates to product evaluations. Because blogs can help an organization tell its story, they can be an important part of content marketing (pages 245–246).

Organizations use blogs to communicate with individuals who seek the less formal, more personal relationships that successful blogs can build. A blog—or even better, a collection of blogs—can help give a human face to an organization. Increasingly, the primary audience for blogs is customers and potential customers. Social Media Examiner reports that almost 70% of organizations use blogs in their social media marketing—and two-thirds of those organizations intend to increase their efforts in that area.

Blogs actively invite written responses, offering a Comments button to allow readers to agree, disagree or extend the discussion.

Ideally, bloggers should write their own posts. Occasionally, public relations practitioners are asked to write posts for others. Though not advisable, this practice is acceptable as long as the purported writer reviews and approves the post.

Blogs exist as webpages but are often reposted to social media such as Facebook and LinkedIn.

### KEY TO SUCCESS

When bloggers represent an organization, they must fully disclose both who they are and the nature of their relationship with the organization. Good blogs are informative, informal and frequently updated. They include links to other information sources.

## Format/Design

Blog entries usually begin with a headline in large, bold type. Beneath the headline, in smaller type, is the date of posting. If more than one writer posts on a particular blog, a line below the posting date specifies the blogger's name. That byline often is a link to more information about the author. That biographical information often is included below the blog post as well. A small photo of the author often accompanies the byline.

Consider including an appropriate photograph or image at or near the top of a post. A growing number of blogs focus on travel, fashions and product demonstrations, making images and links almost indispensable.

The text of a blog is single-spaced, with extra spaces between paragraphs. The text often is a sans-serif typeface such as Lucida Grande.

Blogs contain highlighted links to earlier posts, other blogs, online documents, videos, other websites and/or similar information sources. They also can contain icon/links for sharing to social media platforms such as Facebook and LinkedIn.

Many blogs include subscribe/unsubscribe options at the bottom.

## Content and Organization

The best way to learn to write blogs is to read blogs—as many as possible. Search engines such as Google can easily lead you to blogs in almost any area that interests you.

Employees have been fired for controversial or unapproved blog posts on organizational websites, so if you plan to blog on behalf of an organization, be sure to know the ground rules and to have each post approved before you go online. Ideally, a blog has a strategic purpose that helps fulfill an organizational goal.

Blogs must be updated frequently; many are updated daily.

Successful blogs generally address different aspects of a broad topic, such as politics, product development, fashion or travel. If you're the leader of an organization, almost anything you say about any aspect of your organization has the potential to be interesting. But if you're the head of community relations for a corporation, you might want to confine your posts to discussing community issues and volunteerism. In other words, establish an area of expertise that will keep readers coming back for more.

Blog posts can range from one paragraph to several paragraphs. Be informal: Use the first-person pronouns *I* and *we*.

Blogging guru Mu Lin recommends short paragraphs, frequent internal headlines and use of bullets for lists. Each of those ingredients makes a post easier to read.

Sentence fragments and nonstandard grammar are fine. However, deviations from standard grammar should be intentional; nonstandard grammar should not appear to be the result of sloppiness or ignorance.

Bloggers must frequently review readers' comments and respond when possible: Remember, the goal of most blogs is to build productive relationships. Blogging etiquette allows you to delete obscene or threatening comments but not comments that simply disagree with your post. Many blogs will not post a reader comment until the blogger or a representative has reviewed it.

## BLOG TIPS

1. **Read other blogs**: Not only will you learn more about blogging and particular issues for your own posts, but you'll also discover blogs to which you may wish to link.
2. **Link**: Most blogs have links that allow readers to navigate to other areas of the website or to publicize the blog through social media such as Facebook.
3. **Optimize**: Be sure that your blog is optimized for tablets and phones.
4. **Use keywords**: Keywords can make it easier for search engines to find your blog. Review the search engine optimization tips on pages 19–20.

# Blog Post

## Disconnect!

Bert Mulliner
**May 29, 2023**

So much has changed since we founded MGS Interactive Games in 2006—but one thing hasn't: MGS is a values-driven organization. We frequently discuss the core values that unite our company, and we try to act on them. Our core values are pretty basic:

**MGS Core Values**
MGS Interactive Games was built on three core values:

- Creativity
- Collegiality
- Community

**Living the Values**
Lately, we've been talking about how to act on our top value: creativity. Over the years of creating award-winning games, we've noticed something: Most of us get our best ideas when we're disconnected from technology and are doing some kind of repetitive action. In particular, hiking with our phones put away seems to work a kind of magic for brainstorming new ideas.

In fact, we've learned that when people disconnect from tech for at least 48 hours, they score higher on creativity tests.

So guess what? We're paying MGS employees to get out of the office, disconnect and get into nature. When they come back, we'll keep you posted on their latest ideas. You'll be amazed.

Have you had similar experiences with creativity? Please let me know. We'll share your stories.

COMMENT Subscribe / Unsubscribe

Credit: PhotoMediaGroup/shutterstock.com; Albina Sazheniuk/shutterstock.com

# 2E Podcasts

## Purpose, Audience and Media

A podcast is an audio and/or video file downloaded from the internet. It is typically available as a series, with new episodes acquired either on-demand or automatically through subscription. Video podcasts sometimes are called *vodcasts*, though the term *podcast* applies to both. Podcasts can be an important part of content marketing (pages 245–246). Audiences for podcasts are any public with an interest in your organization. That may include employees, shareholders, journalists, customers and regulators.

The Pew Research Center has reported that more than half of Americans 12 and older have listened to a podcast. That's up from just 18% a decade earlier. One-third of that group had listened to a podcast within the last month and one-in-five within the last week.

As previously mentioned in Section 1F (page 29), podcasts are an inexpensive vehicle through which organizations can extend brand awareness and strengthen relationships. For example, Pew reports that National Public Radio's podcasts, which include *Up First* and *Planet Money*, attracted 7.1 million unique listeners as early as 2018, a 31% increase over the previous year. The benefits NPR derives from these podcasts include enhanced credibility and listeners relying upon it as a trusted source of information.

People can listen to podcasts at hosting websites, including an organization's digital newsroom (pages 132–134). People also can download podcasts to their own computers or to portable devices such as phones and tablets. They also can subscribe to podcasts or can download apps that automatically deliver new programs from favorite sources. Apps such as Apple Podcast also offer access to thousands of podcasts.

**KEY TO SUCCESS**

Success depends on the purpose of the podcast. Although most podcasts are conversational and informal, a successful podcast often begins with a well-written, well-edited script to ensure that the podcast delivers and supports a clear strategic message.

## Format/Design

A podcast can be as informal as a popular musician turning on a recorder and saying whatever comes into their head. But when a podcast represents an organization with specific goals, you should ensure that the final product delivers a clear strategic message. That often means working from a production script.

Audio podcast scripts should resemble audio advertisement scripts (pages 203–209). Video podcast scripts should resemble video advertisement scripts (pages 210–216).

Like audio and video scripts, a podcast script begins with a written title. The title becomes the headline—the webpage link—that people can click on to access the podcast.

Some podcasts feature interviews with guests. In those cases, the host's introductory and closing comments generally are scripted, as are many of the questions. The guest's answers, of course, generally remain unscripted and may be edited for clarity or time considerations.

One caution regarding scripts: They shouldn't sound as if they're being read. To sound natural and conversational, a scripted podcast must be rehearsed. If the individuals featured in your podcast sound unnatural and awkward with a script, consider a less formal approach. You can digitally edit the recorded results to remove awkward pauses and fillers such as *uh* and *y'know*.

Podcasts also allow an organization to present live events for later broadcast. Some of what happens at those events may be unscripted—or at least not scripted by the podcast writer.

When a podcast is hosted by a website, that site should include a podcast's title (the script's title), date of posting and a concise summary. Below the concise summary, consider including any links that can connect listeners to information sources mentioned in the podcast. The summary section also should include a *Comments* link for listener response.

## Content and Organization

Just as with blogs, the best way to learn to write good podcast scripts is to listen to or watch podcasts. Which are your favorites and why? How do they use music and other sound effects? How long are they? Do they feature a single host, or do they include guests in an interview format? For example, *Serial* is one of the most-downloaded podcasts of all time. Over the course of a season, usually 11 episodes, a host guides listeners through a real-life mystery. Each episode lasts approximately one hour and is automatically delivered to subscribers once a week.

New podcasts should be posted frequently. Many organizations update their podcasts weekly. Conventional wisdom holds that unless new podcasts appear at least every two weeks, listeners lose interest. Outdated podcasts fail as sources of current information.

Podcasts often open and close with theme music to help establish a sustained identity. Be sure that you have copyright clearance for any music that you use.

Podcasts can include snippets of sounds from other sources, such as recorded quotations and recordings of live events. Again, be sure that you have written permission to include such snippets.

Many different styles of podcasts exist: informal personal essays (like a conversational blog read aloud), recordings of speeches delivered elsewhere and introduced by the podcast host, recordings of live events introduced by the host, interviews with guests, instructional how-to segments, answers to listeners' questions, collections of short news stories, elaborate documentaries and many other styles. As with advertisements, use a podcast to emphasize the benefits of your message to the audience. Deliver information that can increase your audience's sense of control, companionship or confidence.

The organization's strategic mission and the needs of its audience should determine the style and length of a podcast. For example, many podcasts are timed for individuals who listen while commuting to and from work. The ideal length for a podcast is as many minutes as members of your target audience want the program to continue. Cision reports that the average length of a podcast is 43 minutes. Like a blog, a podcast generally uses informal and conversational language. However, podcast language is never needlessly ungrammatical, which could be distracting to many listeners. The knowledge level of the target audience dictates the level of detail and language used in the podcast.

In writing a podcast script, be sure to follow the "Writing for the Ears" guidelines on pages 29–34.

Some podcasts feature clever, teasing titles such as *The "L" Word* and *Ferry Tales*. Most podcasts, however, lead with more specific titles such as *Fuels of the Future* and *Conversation with CEO Ivy Jones*. Some titles resemble straightforward news headlines; most, however, are concise, specific labels.

Some podcasts lead with a "time and title" marker—something similar to this:

> Welcome back to Tech Talk, a weekly podcast from the software engineers at M-G-S Interactive Games. Today is May 13th, 20–23. I'm M-G-S Software Manager Kris Lancaster, and our topic today is "What's Next in Virtual Reality Games?"

Podcasts often conclude by repeating the podcast's name and by directing attention to the hosting organization:

> You've been listening to Tech Talk, with your host Kris Lancaster. For more information on virtual reality games, please visit our M-G-S Interactive Games website. Until next time—keep playing!

(Note the broadcast-style writing in the previous passages.)

Include a *Comments* link on the website that hosts the podcast. Read and respond to listeners' comments. Those comments may provide ideas for future podcast topics.

Podcasts have the same strengths and weaknesses as the more traditional medium, radio. Both are portable—you can listen to them anywhere. However, like radio, podcasts often reach people who are doing other things, such as driving, cooking, or exercising. Keeping the program short and maintaining a focus on benefits can keep listeners engaged even as they're distracted by other parts of their environment.

## PODCAST TIPS

1. **Brand your podcast**: Don't expect an instant audience for your podcasts. Pick a name that is memorable and resonates with your target audience. Since most people will find your podcast through a search engine, include a keyword or two within the show's name.
2. **Promote your podcast**: Publicize it, particularly through Twitter, Facebook, LinkedIn and other social media. Look for ways to mention the podcasts in other media that you use to reach the target audiences.
3. **Plan your podcast**: Keep a list of show topics. If you have a good idea, write it down. Some podcasts are divided into segments, which can introduce changes of pace and topics that will keep your audience engaged.
4. **Post the transcript**: If the podcast has a transcript, post it online, using search engine optimization techniques (pages 19–20) that make it easy for web browsers such as Google to find it. That heightened visibility may help attract new listeners to your podcast and expand the reach of your message.
5. **Diversify**: Organizations need not limit themselves to one continuing podcast. Organizations could create different continuing podcasts for different audiences, such as investors, employees and customers.
6. **Build your audience**: Create a strong website built around the podcast. This should be part of your digital newsroom or a direct link from it. Encourage audience feedback, using a *Comments* section or inviting listeners to a live chat. By keeping a close eye on analytics, you can learn which topics most interest your listeners.
7. **Archive**: Make all of your podcasts available. As your audience grows, people may want to listen to previous episodes to hear what they missed.

# 2F Websites

## Purpose, Audience and Media

Websites are a primary means of distributing information quickly to a diverse audience. They can inform, market goods and services, share opinions and entertain. Anyone and everyone can have a website.

No other audience is as potentially diverse as a web audience. However, when writing or designing a website, think of a specific user. Give them a name; list their needs, desires and goals. Put yourself in their shoes. Ask yourself these questions:

- To what category of web users (students, parents, alumni, for example) do they belong?
- What is their age?
- What are their interests?
- What language(s) do they speak?
- What do they want to accomplish at the site?
- What are their needs and expectations?
- How technically proficient are they?
- What device will they most often be using—laptop, mobile, tablet, phone?
- How did they arrive at the site—search query, link from another site, etc.?
- At what time of day will they visit the site?
- How long might they stay at the site?
- What will make them return to the site?
- What other websites might interest them?

Web users, from novices to power surfers, do share some distinguishable characteristics. They are busy people who want information quickly. They are impatient. They move through pages randomly and make snap decisions. They don't just read or look at information; they interact with it. They manipulate it.

**KEY TO SUCCESS**

Websites must provide current, content-driven information in an easy-to-navigate format that is relevant to the users' needs.

## Content and Organization

Please see Writing for the Web, pages 19–25 in Section 1.

## Format/Design

So how exactly do users read websites, and what can we learn about webpage design from analysis of the users' behavior? Eyetracking studies in 2006, by Jakob Nielsen, a web usability researcher, show the dominant reading pattern on the web looks somewhat like the letter "F" (www.nngroup.com). First users read horizontally across the upper part of the content area. Next, they move down the page and read a second time horizontally across the page, typically covering a shorter area. Finally they scan down the left side (Figure 2.1). In other words, they read less and less as they move down the page. Nielsen's more recent study also shows that in right-to-left languages, such as Arabic, users flip the F-shape pattern and focus more on the right side of the page.

Early studies by Google refer to this pattern as the Google Golden Triangle—it's the optimal location to place information (Figure 2.2). However, with the rise in mobile usage, more vertical patterns are emerging. Users appear to be consuming content more quickly by scanning vertically down the page (Figure 2.3).

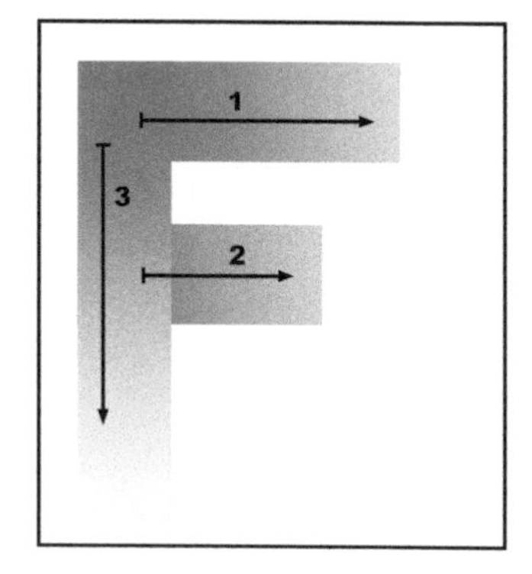


**FIGURE 2.1**
Original eyetracking studies showed that people scan websites much like the letter "F."

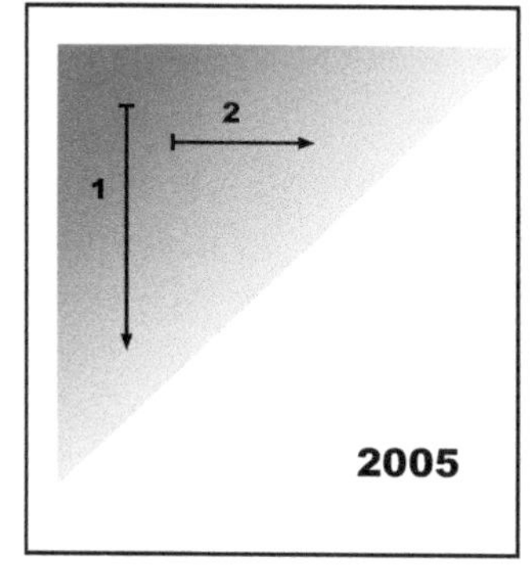


**FIGURE 2.2**
Google's Golden Triangle shows users focused on the upper left portion of a website.

**FIGURE 2.3**
Later studies show a more vertical eye pattern, still placing the emphasis on the left side of the page.

The Nielsen Group recently revisited its original eyetracking research and determined that there are other common scanning patterns in addition to the F-pattern. Some are:

- Layer-cake pattern—Users scan headings and subheadings and skip the underlying content.
- Spotted pattern—Users skip large sections of copy and scan down the page looking for something, such as a link, numbers or specific words or phrases.
- Marking pattern—Users "mark" or focus in one place as they scroll down the page. This occurs more frequently on a mobile website.
- Bypassing pattern—Users skip the first words of a line when all start with the same word, as in some lists.
- Commitment pattern—Users who are motivated or truly interested in your content will read the entire page. This rarely occurs.

However, the F-shaped pattern tends to be the predominant way users view a website when the page includes an abundance of text without breaks, such as subheadings, bullets or boldfaced words. Most web users are looking for quick answers and are not interested in the entire content of your site. Thus, all these scanning behaviors can be bad for business if users are skipping large sections of your web content.

The need for websites to be seamless across all devices also presents design considerations. Responsive design is a web development approach that creates fluid grids that reshuffle page elements depending upon whether the user is on a small or large screen. Responsive design relies on fluid grids designed in percentages to rearrange content and design elements. Consequently, users have different viewing experiences depending upon their choice of device. While the responsive design approach may eliminate the need to develop separate sites for wide-screen monitors, desktops, laptops, tablets and various size phones, it underscores the importance of marrying your website framework with your content and messaging.

So what can we learn from observing website users' behaviors? Writing and designing go hand in hand. Design must enhance readability and not distract from content. The success of the site depends upon how well the viewer can navigate through the site to find specific information. Users rarely think about your site structure. They simply go forward or backward. Users should be able to answer the following for every page on your site:

- Where am I?
- Where have I been?
- Where am I going?
- How do I get there?

## The Anatomy of a Website

Creating or revamping a website for your company or organization is a daunting task that involves much self-examination and awareness, research and data collection, and most of all teamwork. Chances are you'll work with professionals who will guide you through the process.

When beginning the process, ask yourself these two questions:

Question 1: What is the user of my site looking for?

Question 2: What do I want the user to do?

Putting the user's needs first will help you pinpoint your focus throughout the website-building process. You are trying to create a site that will answer a potential user's questions or needs. But just as important is reminding yourself that you want the user to take some action as a result of visiting your site. Your site's success or failure depends upon the answer to this second question.

Every website is unique and should be tailored to the needs of your company or organization. However, the user generally expects to find some standard pages. Here is a list and brief description of some of them.

## Homepage

Your website should have a separate and distinguishable homepage. It is the first page of your website and gives an overview of what your company does. It should function as a front room or lobby that welcomes the visitor. The homepage is where audiences can be split into different interest groups and guided to specific information. For example, a website homepage for your university might have links for current students, prospective students, parents and alumni. Clicking on the appropriate link would send the user to a unique page that addresses their specific interests. The homepage should include a brief explanation of who your company or organization is and what you do, a short description or bulleted list of your products and services, and maybe a brief explanation of the benefits of those products and services.

## About Page

Potential customers want to know more about you. They want to know how and why you do what you do. Here is your chance to share your story—your company history, mission, vision, accomplishments and leaders. This is also the place to point out how you differ from others in your field.

Include visuals to build a personal relationship, such as photos of key personnel, employees in the workplace and/or a short video. Include contact information, including email address, telephone and fax numbers, mailing address, directional map and a contact form with a very obvious Call To Action button.

## Products and Services Pages

Depending upon your business or organization, this section could involve multiple interconnected pages. This is the place where you sell your work product with powerful benefit-driven copy. Start with a brief overview of what you offer, and then list your products and/or services below. Follow this with a description of each product or service. If you have a large number, you might include a link to a landing page to help the user learn more about specific offerings. Be sure to distinguish your company's products and services from your competitors' with facts and details.

## Landing Page

Depending upon the purpose of your website, it may also have one or more landing pages that share the same domain. Landing pages differ from the homepage in several ways.

| **Homepage** | **vs** | **Landing page** |
| --- | --- | --- |
| Introduces site and directs user to internal content | | Encourages user to click on a single Call To Action button |
| Offers a broad content overview | | Makes a specific offer |
| Uses navigation as a vehicle to go deeper into site | | Uses no links or other navigation tools |
| Positions and describes your overall business | | Drives user to a desired action |

The goal of the landing page is to drive traffic for a specific marketing campaign. Thus, landing pages will change depending upon the campaign. Your landing page appears when a user clicks on your ad or a search engine result link about that campaign. The purpose is to convert—whether that means generating leads, promoting phone calls, increasing signups or initiating an online chat with a potential customer.

### LANDING PAGE DESIGN AND COPY TIPS

1. **Use large font sizes**: These grab attention and enhance readability.
2. **Use NO navigation tools**: Your goal is to keep the user on your page as long as possible or until they convert.
3. **Use only one visual**: Usually the visual is a picture of the product or service being used.
4. **Write a clear, simple headline**: Focus on the offer. Keep it simple and repeat words that were used in the initial ad or search engine results.
5. **Write sales-driven copy**: Focus on benefits rather than features.
6. **Use action words**: Focus on value and urgency with words such as *reserve, limited time only* or *FREE*.
7. **Offer a guarantee**: Give the user a safety net to try your product or service.
8. **Provide social proof**: Include testimonials, reviews, social media posts or media mentions (*As seen on . . .*) to show that others use and trust your product.
9. **Make it easy to act**: Make the Call To Action button highly visible with use of color, shape or other visual cues that direct the eye. In other words, the Click-Through button should stand out.
10. **Use a simple form**: In requesting information, ask only for information you need to follow up effectively. Your landing page success will be measured by the percentage of users who complete and submit the action form.
11. **Keep selling**: Use a thank-you page to confirm a successful signup.

## FAQs Page

This page or pages includes questions frequently asked by customers. The FAQ page saves time for both you and the customer because you've provided a place that is easily accessible for customers to find answers or to relieve their concerns. Choose questions that focus on opportunities to remove barriers to action or sales. Use a casual, friendly tone, and keep answers short and to the point. Don't shy away from difficult questions and ALWAYS, ALWAYS be honest.

## Testimonials and Reviews

Testimonials help build credibility and trust. They encourage others to buy, sign up or use whatever product or service you are promoting. These pages are often the most visited pages on your site. Testimonials come directly from customers who have used your product or service. They may include customer experiences, case studies, photographs or direct quotations.

When getting a testimonial from a satisfied customer, ask to use their real name. Real endorsements from recognizable businesses and organizations carry more weight than do anonymous ones.

## Contact Page

This page shows potential customers a variety of ways they can contact you. You should include telephone and fax numbers, email address, mailing address, social media accounts and a clickable "Email Us" or contact form. Consider putting a telephone number for sales inquiries in the footer of every page of your website.

Businesses with a brick-and-mortar location should also include a map with a link to Google Maps. Initial searches often start with a request for a business in a specific town or at a GPS location "near me."

## Blog

A company or organization blog may be the most affordable and greatest marketing tool available. A blog is a series of posts about related topics usually listed in reverse chronological order, with the most recent blog post appearing first.

To get the maximum search engine optimization benefit, list your blog under your company domain (www.yourcompany/blog). Search engines, such as Google, will index pages to your blog just as they do other pages in your website. Each post you add increases your reach and could draw more potential customers to your site.

Blogs also produce content that can be shared or liked by potential customers on social media, such as Facebook and Twitter. They also establish your company as an "expert" in your field that freely shares knowledge with others.

Consider offering a freebie to users who add their name and email address to the blog's mailing list. Also add a Call To Action button to encourage users to subscribe to your blog.

## Media Page

A media page, also known as a newsroom or digital newsroom, makes it easy for members of the media to contact you and to obtain accurate information about your company or organization for publication. First and foremost, include contact information on how you prefer the media to contact you (email, telephone number, etc.) Also include backgrounders and fact sheets (see pages 135–141), short bios of key personnel with downloadable headshots, awards or accolades your company has won, links to articles and/or blogs that are about or mention your company and news releases. Also include a zip file with a complete media kit (see page 132).

## Privacy Policy

Your privacy policy lets your users know that their personal information is safe with you. It explains how information that is collected is used, whether or not it is shared with others and, if so, with whom. Boilerplate privacy policy statements are available online.

## Terms and Conditions

This page spells out the rules for using the website. Key issues such as acceptable use, privacy, cookies, registration and passwords, intellectual property rights, links to other sites, etc. are explained. Again, boilerplate privacy policy statements are available online.

## Sitemap

The sitemap is where you provide information about the pages, videos and other files on your website and the relationships among them. Search engines, such as Google, read this file and gather valuable information about your website. Unless your site contains only a few pages, a well-designed sitemap will make it easier for Google and other search engine crawlers to understand and index your content. Go to support.google.com for information on how to build and submit a sitemap.

Other pages that might be appropriate to your website include jobs and careers, support and forums, portfolios, "404 error" (a page that no longer exists), search result page, events page and others.

All the pages mentioned earlier follow some basic guidelines for copy and design. As discussed earlier, eyetracking research indicates that the most important location on a webpage is the upper left-hand corner. This is where users first look. Therefore, you should begin each page with a strong headline followed by opening sentences that deliver the most important information. Subheadings and bullets that begin on the left-hand side of the page should start with information-loaded words or keywords that boost SEO (see pages 20–22). Since we know that users do not read word-for-word, the site should include visual elements, such as subheadings and bullet points, to help the user scan down the page to easily extract the desired information.

Users prefer to scroll rather than click. Scrolling is quicker than clicking, especially for those on a mobile device. Scrolling is not decision based. Users simply scroll without having to make a decision to leave the page. Therefore, it's important to give the user a reason to scroll with good content and strong visuals.

For more information on writing for the web, see pages 19–25.

## WEBSITE FORMAT/DESIGN TIPS

1. **Make site speed an absolute priority**: If your site is slow to load, users will not stay.
2. **Make your site mobile friendly**: More than 50% of web traffic comes from mobile devices. Mobile responsiveness refers to how well your site appears on small-screen devices. Consider designing the mobile-friendly version first.
3. **Simplify navigation**: Decrease the number of options and the amount of clutter. Phase out sidebars, sliders, carousels and slideshows.
4. **Label buttons clearly to aid in navigation**: For example, *About Us* is more descriptive than *Overview*.
5. **Respect the fold**: Placing key information above the fold is still a good principle whether in print or online. What is at the top of the screen determines whether the user will scroll down to learn more. It's important to remember, however, that good content should continue throughout the page.
6. **Don't clutter your website with unnecessary graphics**: Less is more when it comes to web design.
7. **Use the basic principles of design to arrange elements on your page**: Balance/visual weight, movement/repetition, emphasis, contrast, proportion, space and unity/alignment/proximity (see pages 35–41). Surround your most important elements with white or negative space. Use design to direct the user's eyes to the most important elements on the page.
8. **Choose a color palette (a group of related, complementary colors) to use throughout your website**: Select tones within that particular palette. Complementary colors, those that lie 180 degrees across from each other on the color wheel, provide the most contrast. Poorly used color can create image overload that distracts from the content and reduces text clarity (see pages 38–41).
9. **Choose graphics that work well in black and white as well as in color**: Some colors are difficult for the visually impaired to see, so steer away from yellows and grays.
10. **Use photography showing real people to set the appropriate tone and mood for your site**: Avoid stock photography, if possible, and be sure you know the source of your graphics and potential copyright limitations before using them. Not everything you find is yours to use free of charge.
11. **Use video if appropriate**: SEO technology loves videos, which can mean better search engine results for your website. Users have the option to click and learn more.
12. **Choose typography that complements your brand**: Use only verified *web-safe fonts* that can be displayed on most devices. Be consistent in your use of size, style and color with respect to headlines, subheadings and body text throughout your site. Avoid cursive or showy fonts for body text.

13. **Keep line length short**: The normal reading distance the eyes can span is only three inches wide or approximately 12 words. Don't write paragraphs longer than three to four lines. Use columns, tables and graphics to narrow the line length. Sans-serif fonts and exaggerated x-heights are easier to read on screens (x-height refers to the height of a lower-case x within a typeface).
14. **Include social media sharing buttons**: Facebook, Instagram, LinkedIn, Google+, Twitter, YouTube and Pinterest are the most common. Choose which buttons to keep and which to scrap depending on your website. Consider placing them in the top-left side of the header or above or below posts.
15. **Study websites that *you* use and like**: Without duplicating their specific words and images, consider incorporating their successful strategies into your website.

# 2G News Release Guidelines

## Purpose, Audience and Media

A news release is a document that conveys newsworthy information about your organization to the news media. Journalists and bloggers generally agree that a newsworthy story has at least one (and probably more) of the following elements:

- **Timeliness**: The story contains new information.
- **Impact**: The story affects media readers, viewers or listeners.
- **Uniqueness**: The story is different from similar stories.
- **Conflict**: The story involves a clash of people and/or forces, such as nature.
- **Proximity**: The story describes events geographically close to the target readers, viewers or listeners.
- **Celebrity**: The story involves a famous person, such as a politician, business leader or entertainer.

Traditional news releases are written as ready-to-publish news stories. You write a news release in the hope that journalists and/or bloggers will take its information and publish or broadcast it, thus sending your news to hundreds or thousands—perhaps even millions—of people.

The most significant difference between an organization's news release and a journalist's news story is perspective: A news release is written from the organization's point-of-view, emphasizing the values and information it considers most important while presenting the message it wants to deliver. However, the news release must be complete, fair and objective; it should sound as if a good reporter wrote it.

Don't be hurt if your news release isn't published or broadcast verbatim. Most are not. Journalists and bloggers often use news releases as story tips, and they rewrite your work, sometimes with additional information. If their stories and posts include your main points and don't introduce any negatives, your news release succeeded.

The news release is often called the press release, a term that is outdated. Increasingly, we get our news not from print media (which use a printing press) but from online sources, television and radio. The term *news release* seems more appropriate.

Still, traditional news releases generally are written as if they were for newspapers. Other news media, such as television stations, then edit news releases for their particular needs.

In general, three styles of news releases exist, each of which is described in the following sections:

1. Announcement (the straight news story)
2. Feature story (a combination of information and entertainment)
3. Promotional (pure promotional writing generally meant for blogs and newsletters)

Two other documents are similar to news releases. Unlike news releases, they are not created in ready-to-publish formats. Each of these is described in upcoming segments:

1. Media advisories (quick facts on breaking news stories)
2. Pitches (exclusive offers of stories to particular journalists)

The audience of a traditional announcement news release is a journalist. To be a successful news release writer, you must focus intensely on what journalists like (and avoid what they dislike) in news stories. They like conciseness; they dislike wordiness. They like specifics; they dislike generalities. They like reputable sources; they dislike unattributed opinions and unattributed claims. They like objective facts; they dislike promotional writing. They like honesty and candor; they dislike dishonesty and evasion. Journalists have a time-honored place for announcement news releases that violate these guidelines: the wastebasket, digital or literal. (Promotional news releases, which generally target bloggers and informal news outlets, violate some of these norms.)

News releases are among the oldest documents in all of public relations. With all the changes in news media, do journalists still really want them? The answer is a resounding yes. Modern research shows that almost 90% of journalists find news releases valuable.

News releases exist in a variety of media: Surveys show that journalists prefer email delivery. News releases also can be placed in digital newsrooms (pages 132–134) on organizational websites. For live news conferences and trade shows, news releases can appear on paper or on USB flash drives for on-site distribution. News releases have even been written on the labels of champagne bottles and sent to journalists (attached to a full bottle, of course).

## KEY TO SUCCESS

A traditional announcement news release should contain only newsworthy information. It should not be a thinly disguised advertisement for your organization. A good news release has a local angle; that is, journalists and bloggers read it and quickly see that the information it contains is relevant to their readers, listeners or viewers.

# Format/Design

## *Email*

Surveys show that more than 90% of journalists prefer to receive news releases via email, as opposed to links in texts and tweets, phone calls or old-fashioned snail-mail paper documents. The same surveys show that journalists dislike attachments. They prefer emails in which the news release is the on-screen message, ideally with an embedded image and links.

Email news releases generally have templates that include organizational letterhead. Such templates can help reporters immediately identify the source of the news release. Email news releases can and should contain links to additional information about key organizations and individuals.

## *PDF/Paper*

PDF news releases often exist in an organization's digital newsroom and on flash drives for distribution at trade shows and other events that may involve more than one news release or newsworthy document. As a PDF or paper document, a news release should be on your organization's stationery. Use letterhead stationery (with your organization's logo, for example) only for the first page. If the news release extends to a second page, however, don't switch to a different color or quality of paper with paper news releases. Some organizations use special news release stationery that clearly labels the document as a news release.

## *Headings and Contact Information*

### *Email*

In email news releases, the all-important subject line comes first. Type the key fact of your news release in the subject line. A good subject line is newsworthy, specific and concise; it shows journalists that the related email contains news of interest to their audiences. Subject lines usually are more concise even than headlines.

The email news release itself has fewer headings than a PDF or paper news release. Before the headline, type only "For Immediate Release." Below that, type the date on which you send the release. (Unless there is prior agreement on not publishing a news release until a specific time, known as an *embargo*—see page 102—always assume that the news will go public once it is released.) Unlike PDF/paper news releases, email news releases include the "For More Information" data below the story. At that point, include the contact person's name, title, phone number and an email address/link. (The contact person is your organization's official spokesperson for this news release.)

Both the "For Immediate Release" at the top and the "For More Information" at the bottom are aligned on the left margin.

Leave a blank line between the headings and the headline. The headings of an email news release should look something like Figure 2.4.

To: JaneQReporter@newspaper.com

Subject: Circle City Blood Drive Scheduled

For Immediate Release
Nov. 20, 2023

**Circle City Red Cross schedules downtown blood drive**

**FIGURE 2.4**

## *PDF/Paper*

If your digital template or paper stationery doesn't label the document as a news release, type "News Release" in big, bold letters—usually 24-point type. Below that, begin the actual news release, in 12-point type, with headings that specify "FOR IMMEDIATE RELEASE," the composition date and "FOR MORE INFORMATION" data: a contact person, the person's title, a phone number and an email address. (The tradition is to use uppercase letters for those two phrases in PDF or paper news releases and lowercase letters in emails.)

The headings should be single-spaced.

***ORGANIZATION LETTERHEAD***

**News Release**

**FOR IMMEDIATE RELEASE**
Nov. 20, 2023

**FOR MORE INFORMATION:**
Catherine Jones
Public Relations Director
(555) 123-4567
cjones@xyz.org

**Circle City Red Cross schedules downtown blood drive**

**FIGURE 2.5**

Leave at least an inch between the headings and the headline. All together, the headings of a PDF/paper news release should look something like Figure 2.5.

## The Headline

### Email and PDF/Paper

Your headline should be a newspaper-style headline (see page 99). Boldface the headline. Capitalize the first word and any names (of people, buildings, organizations and so on). Lowercase all other words, just as most newspapers do. Keep it short: Average headline length for news stories is six words—no more than 60 characters and spaces.

## The Text

### Email and PDF/Paper

Single-space the text of news releases. Do not indent paragraphs; instead, include a blank line between paragraphs. Some old-fashioned paper news releases still double-space the text, but that practice dates back to when journalists used pencils to edit the document and wanted that extra space. Very few modern news releases are double-spaced.

The text of a news release should be long enough to tell the story concisely—and no longer. The entire release rarely should be more than two pages—one front and one back or two separate pages. Make it shorter, if possible. Many news releases are one page.

## Other Format Traditions

### Email

After the text of an email news release—but before the contact information—type "-30-" or "###." Include a blank line between the end of the text and that symbol.

At the bottom of an email news release, include subscribe/unsubscribe links that allow recipients to sign up to receive your news releases and to cancel that service. Include social media icons/links, such as those of Facebook and Twitter, that allow recipients to share your news release with others.

### PDF/Paper

If the news release is more than one page, type "-more-" or "-over-" at the bottom of each appropriate page. Beginning with the second page, place a condensed version of the release's headline (called a "slug") and the page number in the upper-right corner. After the last line of the news release, space down one more line and type "-30-" or "###."

Staple the pages of a paper news release together. Never trust a paper clip.

## Content and Organization

In all news releases—email, PDF and otherwise—focus on your audience: journalists and bloggers who seeks newsworthy information for their audiences. What kind of information is newsworthy?

- Timely information that affects members of a news medium's audience. Such information is said to have "local interest"—an important quality to journalists.
- Timely information that is unusual or exceptional
- Timely information about a well-known individual or organization
- Survey research shows that journalists particularly like new, interesting and useful (to their audiences) information, particularly relevant surveys and reports. They also like news about new products that definitely will appeal to their audiences.
- For additional qualities of newsworthiness, see page 94.

### *The Subject Line*

The subject line of an email news release is hugely important. Research shows that 80% of journalists won't open an email that doesn't quickly, clearly and specifically promise new, relevant news. Subject lines usually are even more concise than headlines.

### *The Headline*

News release headlines are written in newspaper style. Most newspaper headlines are, roughly, complete sentences.

Most newspaper headlines are written in present tense, which, in headline grammar, means recent past tense. For example, "Google celebrates anniversary" means that Google celebrated recently—probably earlier today or yesterday. Some headlines, however, require future tense. If Google is planning an anniversary celebration and you are writing a news release to publicize that, the headline would be "Google to celebrate anniversary."

A good headline includes local interest and summarizes the story's main point. Whenever gracefully and logically possible, mention your organization's name or product in the headline.

### *The Dateline*

The text of a traditional news release begins with a dateline in capital letters and a dash (for example, "DALLAS—"). Datelines give the location of the story. They help establish local interest and answer the reporter's question *where?* Datelines also can include dates (for example, "DALLAS, Jan. 24—"). Feature news releases often lack datelines. The Associated Press Stylebook can tell you which city names need to be followed by an abbreviated state name in a dateline.

## The Text

With or without a dateline, the first sentence of a traditional news release should establish local interest and move right to the news. A good newsworthy first sentence often concisely covers *who, what, when* and *where*. In a traditional news story, the first paragraph, also called the lead, includes the most important information about the story. It never relies on the headline to supply information. Instead, the headline summarizes information included in the lead.

Research shows that you have three seconds to capture the journalist's attention and persuade them to keep reading: The first sentence must immediately present important news or, in a feature news release, immediately offer an irresistible hook.

A traditional news release is structured as an inverted pyramid, which means that the most important information is at the top of the story (the widest part of the upside-down pyramid). As the story continues and the pyramid becomes narrower, the information becomes less and less important.

Include a pertinent, attention-grabbing quotation from a representative of your organization in the second or third paragraph. Such quotations can enliven news releases, making them more attractive to journalists and news audiences alike. Good quotations provide color, emotion or opinion—or all three. However, journalists frequently complain about boring quotations within news releases: Surveys show that 75% of journalists tend to delete those quotations because they're bland and pompous. Avoid quotations that recite sterile facts or statistics or that do nothing but cheerlead.

Survey research shows that more than 80% of journalists prefer emails with links to additional information.

## Visuals

Journalists prefer multimedia news releases, particularly those with embedded photographs (avoid attachments): More than three-fourths of journalists say they prefer news releases that include a downloadable, high-resolution image. Research from Business Wire, a news release distribution company, found that multimedia news releases are three times more likely to capture and hold journalists' interest than are plain text releases.

## Optional Notes to the Editor

If some information, such as the spelling of a name, is unusual, include a "Note to the Editor" after the "-30-" or "###" to inform editors that your information is correct. This is not necessary for routine information.

## *Distribution and Follow-Up*

Send news releases to specific journalists and bloggers. Companies such as Cision offer databases of contact information for journalists ranging from local to international. Some local chambers of commerce also can supply local media lists to organizations.

If the news release announces an event, be sure that newspapers, radio and television stations and news websites receive it about 10 days before the event. Magazines generally need more advance notice than that; six months isn't too early for some monthly or quarterly magazines.

Consider paying a distribution service, such as PR Newswire or PRWeb, to distribute your news release. Such services can distribute your release to appropriate media throughout the world. These services also can archive your news releases and enhance them with search engine optimization (SEO) technology to highlight keywords, which increases the odds that journalists doing online searches will find the news release. News releases archived on your organization's digital newsroom also should be enhanced with SEO technology (pages 19–22).

Send only one copy of your news release to each appropriate news medium. Don't send your news release to several reporters at the same news outlet, hoping it will interest one of them. Send it to the most appropriate reporter or editor at each news medium.

A continuing debate in public relations concerns follow-up calls—that is, telephoning journalists to ensure that they received the news release and to offer assistance. Most journalists resent such calls unless you've offered an exclusive story (see pages 121–122). If your news release presents a story of interest to a journalist's audience, the journalist will call you if they need more information. Journalists are too busy and receive far too many news releases each day to answer questions from eager news release writers. However, some situations do justify a follow-up call. For example, if you email or text a media advisory (see pages 115–117) on a fast-breaking news story that you know journalists will want to cover, they probably will appreciate a quick follow-up call to ensure they received the information.

If you must call regarding a news release of crucial importance, be polite. Remember: Journalists and bloggers are under no obligation to use your story. If you are unable to reach the journalist or blogger, don't leave more than two messages. Don't earn a label as a nuisance.

Never ask journalists or bloggers if they used your news releases. That tells them that you're not reading, listening to or watching their work. (And if you aren't, you should be!)

## NEWS RELEASE TIPS

1. **Be available**: Ensure that the phone number in the For More Information heading provides 24/7 access to the contact person.
2. **Save the date**: Avoid using the words *today, yesterday* and *tomorrow* in your news release. Journalists and bloggers almost always will have to change those words. For example, your *today* probably will be incorrect by the time your news release is published or broadcast. Using an actual date—for example, Jan. 23—can solve this problem. Daily news media, such as digital and paper newspapers, often use days of the week, as in "XYZ Partners will build two factories in Puerto Rico in 2023, the corporation announced Wednesday." The present-perfect verb tense can be used to denote the immediate past, as in "XYZ Partners *has announced* third-quarter profits of $50 million."
3. **Use *said***: Use past-tense verbs to attribute quotes. Use *said* instead of *says* in print-oriented news releases.
4. **Be precise and concise**: Every word that journalists print, post or broadcast costs money. Be clear, and don't waste words.
5. **Focus on local interest**: Ask yourself why your news will appeal to the audience of each medium that will receive your news release. Use the news release's headline and first sentence to spotlight local interest. (The word *local* doesn't have to mean hometown. For example, a news release about an important new product that will be used by consumers throughout the world has "local" interest to readers everywhere. Finding individual hometown angles, however, can strengthen a news release.)
6. **Avoid promotional writing**: You, as the writer, should be objective. Don't include unattributed opinions (your opinions) or unattributed claims about your organization's excellence.
7. **Edit and review**: Be sure that your supervisor and/or your client review the news release before it is distributed. After you review their comments or suggested revisions, you may need to remind them that a news release generally is an objective, unbiased news story.
8. **Link**: Email news releases can and should contain links to additional information about key organizations and individuals.
9. **Don't embargo**: Avoid so-called embargoed news releases—that is, news releases that aren't for immediate publication. With an embargoed news release, you ask the editor to hold the information until a specified release date. Don't make a practice of asking media outlets to delay the publication or broadcast of newsworthy stories. Embargoes generally work only when journalists and strategic writers agree in advance that a situation merits special treatment.
10. **Include a boilerplate**: Some news releases conclude with a "boilerplate" paragraph or passage—a standard brief biography of the organization. Boilerplates can add additional, nonessential "who" information. The final paragraph of the news release on page 105 is a boilerplate.

# 2H Announcement News Releases

## Purpose

The announcement is by far the most common type of news release. Use announcement news releases for standard "hard-news" stories—for example, the announcement of the opening of a new production facility. (Please review the general guidelines for news releases, pages 94–102.)

## Format/Design

Follow the general guidelines for news releases, pages 94–102.

## Content and Organization

The announcement news release is a straightforward, objective news story. It begins with a traditional news headline (page 99). The lead (the opening paragraph) covers the most important aspects of *who, what, where, when, why* and *how*. In fact, the ideal first sentence of an announcement news release often specifies *who, what* and *when*—with *where* often being in the dateline. For example, in the first sentence of the news release on page 105, we have *who* (Bert Mulliner and MGS Interactive Games), *what* (the upcoming game in a popular series has been named and tested), *when* (May 29) and *where* (Austin, Texas).

The entire news story follows the inverted pyramid structure (page 100); in other words, the information in the story becomes progressively less important. The least important (but still newsworthy) information comes last in an announcement news release. That final paragraph may be a boilerplate (page 102).

Announcement news releases usually are written in past tense. If, in your lead, you need to establish that your news happened in the very recent past, you can use present-perfect tense (using a form of *to have*)—for example, "XYZ Partners has announced third-quarter profits of $50 million."

An announcement news release often includes relevant, newsworthy quotations from appropriate sources, such as members of your organization's management team. (Please review the cautions about quotations on page 100.) Trade magazines, business journals and small newspapers sometimes reprint announcement new releases verbatim. Other media may turn them into brief announcements or use them to generate longer stories. Remember: The news media may ignore a news release altogether, especially if it's poorly written, too promotional or lacks local interest.

# Announcement News Release (Email)

To: JaneQReporter@VPlayToday.com

Subject: Starklight #5 named and tested

## News Release

**For Immediate Release**
**May 29, 2023**

**MGS Names and Tests New Starklight Game**

AUSTIN, Texas – The title of Starklight #5 will be The Dulwich Quest, MGS Interactive Games President Bert Mulliner announced on May 29. The Dulwich Quest is the fifth game in the Starklight series, which features games that pit individual players against programmed competitors or live players. The Dulwich Quest will be available for purchase on Sept. 14.

"We should have waited to announce this, but I'm sick of all the secrecy," Mulliner said. "We're getting upward of 300 emails and texts a day, all asking what the name will be and if we've done any test tournaments. Starklight fans are so loyal that it just seemed cruel not to share what we know."

Gamers test The Dulwich Quest, the newest entry in MGS's Starklight series. MGS tested the new game at four undisclosed locations on May 25. Gamers were sworn to secrecy until the game's debut in September.

Mulliner said that MGS tested The Dulwich Quest on May 25 at four undisclosed locations in the United States. Gamers in each location signed pledges of secrecy in return for free copies of the game when it debuts in September. Mulliner said that the gamers, whose identities MGS will not reveal, gave the new game rave reviews and offered three suggestions for improvement that MGS designers are now testing.

The Starklight series debuted in 2005 with Starklight #1: Pelican Abbey. According to sales figures from VPlayToday magazine, the Starklight series is the second-bestselling video game series in history, trailing only Quaxart's Jupiter937 series. Suggested retail price of The Dulwich Quest is $60.

Mulliner founded MGS Interactive Games in Austin, Texas, in 2002. The company has 430 employees and, in 2018, had revenues of $840 million.

###

For More Information:
Catherine Jones
Public Relations Director
(555) 123-4567
cjones@mgsintgames

Subscribe / Unsubscribe

Credit: Ivanko80/shutterstock.com

## Announcement News Release (PDF/Paper)

2010 Ridglea Drive • Austin, TX 55111 • 555-999-5555

# News Release

**FOR IMMEDIATE RELEASE**
May 29, 2023

**FOR MORE INFORMATION:**
Catherine Jones
Public Relations Director
(555) 123-4567
cjones@mgsintgames.com

**MGS Names and Tests New Starklight Game**

AUSTIN, Texas – The title of Starklight #5 will be The Dulwich Quest, MGS Interactive Games President Bert Mulliner announced on May 29. The Dulwich Quest is the fifth game in the Starklight series, which features games that pit individual players against programmed competitors or live players. The Dulwich Quest will be available for purchase on Sept. 14.

"We should have waited to announce this, but I'm sick of all the secrecy," Mulliner said. "We're getting upward of 300 emails and texts a day, all asking what the name will be and if we've done any test tournaments. Starklight fans are so loyal that it just seemed cruel not to share what we know."

Mulliner said that MGS tested The Dulwich Quest on May 25 at four undisclosed locations in the United States. Gamers in each location signed pledges of secrecy in return for free copies of the game when it debuts in September. Mulliner said that the gamers, whose identities MGS will not reveal, gave the new game rave reviews and offered three suggestions for improvement that MGS designers are now testing.

Gamers test The Dulwich Quest, the newest entry in MGS's Starklight series. MGS tested the new game at four undisclosed locations on May 25. Gamers were sworn to secrecy until the game's debut in September.

The Starklight series debuted in 2005 with Starklight #1: Pelican Abbey. According to sales figures from VPlayToday magazine, the Starklight series is the second-bestselling video game series in history, trailing only Quaxart's Jupiter937 series. Suggested retail price of The Dulwich Quest is $60.

Mulliner founded MGS Interactive Games in Austin, Texas, in 2002. The company has 430 employees and, in 2018, had revenues of $840 million.

###

# 21 Feature News Releases

## Purpose

Feature news releases focus on "softer," less important and less immediate news than do announcement news releases. Feature news releases often are human-interest stories that highlight some aspect of your organization. Feature news releases are not as common as announcement news releases and traditionally attract less media attention. Always consider whether your feature news release could work as an announcement news release. However, if your organization has an interesting story, but it's not a traditional hard-news story, you should consider a feature news release or a pitch (see pages 118–123). (Please review the general guidelines for news releases, pages 94–102.)

## Format/Design

Follow the general guidelines for news releases (pages 94–102).

## Content and Organization

Feature news releases often present entertaining human-interest stories, such as the efforts of an officer of your organization to hire the homeless. Other feature news releases focus on topics bigger than your organization and use representatives of your organization as experts.

For example, Hallmark Cards writes feature news releases on the history and traditions of important holidays, such as Mother's Day. In addressing these interesting topics, the news release uses Hallmark experts and research for evidence, thus bringing credibility to the company and linking it to holiday traditions. Some feature news releases include information from nonemployee, noncompetitive sources to round out the story.

Regardless of the topic, a feature news release also should reflect the organization's underlying key messages and values. For example, a story about a sponsored charity event carries with it an unspoken message of the organization's commitment to public well-being. Whether to promote good employee relations or raise public awareness of an issue important to the organization, every news release, including feature news releases, has a purpose.

Avoid the temptation to include unattributed opinions in feature news releases. Like all news releases, features must be objective and unbiased. Good feature news releases sound as if they were written by journalists, not by promotional writers.

Feature news releases generally begin with a traditional news headline (see page 99). However, many feature news releases attempt to include clever wordplay, such as a pun, in the headline.

Feature news releases don't have traditional news leads. Instead, the lead attempts to spark the reader's interest with a question, an anecdote, an image or a similar device.

Feature news releases use storytelling skills, so they're not inverted pyramids, as are announcement news releases. The most dramatic paragraph in a feature news release might be the final paragraph.

Unlike announcement releases, feature news releases often use present tense to attribute quotations—for example, *says* instead of *said*. Present-tense attributions can help create the sense that a story, not just a report, is being told. (Check to see which verb tense your target medium tends to use in similar stories.)

Because feature news releases can help an organization tell its story in an interesting, engaging manner, they can be an important part of content marketing (pages 245–246) and of digital newsrooms (pages 132–134).

A more extensive discussion of feature stories can be found in Newsletter and Magazine Stories, pages 145–153.

# Feature News Release (Email)

**To:** JaneQReporter@newspaper.com

**Subject:** Duck race for local Donors United

## News Release

**For Immediate Release**
**April 23, 2023**

**MGS to sponsor duck race for Donors United**

AUSTIN, Texas –MGS Interactive Games intern Andrea Smith has hatched a ducky idea for raising money.

"It started in my bathtub," she says. "But that's probably more than people want to know. Let's just say that I was playing with two rubber ducks and got a weird idea."

MGS will sponsor a rubber duck race on Patterson Creek on June 7.

Smith's weird idea may help feather the nest of the Travis County chapter of Donors United. On Friday, June 7, MGS will sponsor a rubber duck race on Patterson Creek. Five bucks buys two ducks, and the first duck to float to Old Bridge wins its owner $100. The race begins at noon at McDaniel Park. Anyone age 18 or older may enter.

"We were a little surprised when Andrea came to us with the idea," says Jane Evers, director of the Travis County chapter of Donors United. "But she's put together a great program."

Ducks go on sale Monday, May 5, at MGS headquarters at 2010 Ridglea Dr. Smith is a sophomore at Palmquist University in Austin.

"MGS is delighted to help the Donors United in this manner," says MGS President Bert Mulliner. "Especially if one of my ducks wins."

###

For More Information:
Catherine Mallard
Public Relations Director
(555) 123-4567
cjones@mgsintgames.com

Subscribe / Unsubscribe

Credit: Dale Berman/shutterstock.com

# 2J Promotional News Releases

## Purpose, Audience and Media

Unlike a traditional announcement news release (pages 103–106), a promotional news release is *not* an objective telling of a news story. Instead, just as its name suggests, a promotional news release openly tries to generate excitement and media buzz about the news it contains: It uses promotional writing, which is a no-no in announcement news releases. Announcement news releases deliver the plain facts of a story in an objective tone. In strong contrast, when singer-songwriter Meghan Trainor released a collection of love songs for Valentine's Day 2019, the accompanying promotional news release was so sexually explicit that the news release itself made news: The staff of the pop culture magazine *Paper* even featured a story on the news release itself and declared, "None of us have been okay" since reading the over-the-top story. Of course, *Paper* shared the entire promotional news release verbatim.

For Trainor and her PR staff, that meant Mission Accomplished.

The audience for a promotional news release, therefore, is *not* mainstream, objective journalists who want facts and unbiased reporting about an important news story. Instead, the audience for promotional news releases includes bloggers and social media influencers. The audience also includes editors of digital newsletters that focus on well-defined areas such as music, movies or video games. All of these audiences seek interesting, fun and relevant stories to share with their own audiences. For example, when the digital newsroom within the website of Marvel Entertainment (home of Captain America, Captain Marvel, Black Panther and other superheroes) featured a story titled "*Iron Man 3* Is a Christmas Movie—Here's Why," that story was republished verbatim in dozens of blogs, newsletters and entertainment websites. It was free, relevant content for those sources, and it was great publicity for Iron Man and Marvel.

Promotional news releases are multimedia, containing words, links, images and, sometimes, videos. They tend to use three types of media for distribution: (1) an organization's own digital newsroom; (2) news release distribution companies such as PR Newswire or PRWeb that create highly targeted email lists to reach news sources that would want such information; and/or (3) emails to logical recipients, such as entertainment bloggers. Digital newsrooms also often allow journalists to sign up to automatically receive timely news releases, usually via email. Organizations also use social media to draw people to the promotional stories in their digital newsrooms.

**KEY TO SUCCESS**

A promotional news release tells a multimedia, newsworthy story (a story that often includes links, visuals and the basics of who, what, where, when, why and how) in an entertaining, subjective voice.

## Format/Design

A promotional news release, whether an email or a webpage within a digital newsroom, usually begins with a headline and a large image. A subheadline sometimes follows the headline.

Promotional news releases are single-spaced. Additional images, videos and links may be placed within the text of the story.

Promotional news releases often are four to eight short paragraphs. Some are shorter and some are longer, depending on the needs of the specific story. Most don't have indented paragraphs. Instead, they include an extra line of space between paragraphs.

Social media icons, such as those of Facebook and Twitter, often appear at the bottom of the release, allowing readers to share the story through their own social media accounts.

## Content and Organization

Most promotional news releases don't score as high on the OMG scale as the infamous Meghan Trainor release. Instead, most deliver basic information (who, what, where, when, why, how) in an informal, chatty, opinionated tone—a promotional tone. The content of a promotional news release is goal-oriented: When strategic writers understand the goal of a specific news release, they will know what information to include in the story. Promotional news releases should not be used for important legal announcements, such as the hiring of a new CEO, a product recall or other matters for which an informal, subjective approach would be inappropriate.

Just as with an organization's social media posts, a voice book for promotional news releases can be important. A voice book specifies the personality of the voice within a document or series of documents. Consider specifying three adjectives to describe the voice within your organization's promotional news releases—for example, informal, funny and edgy. A voice book also can guide decisions on matters such as the use of profanity, sexual explicitness and more.

As with so many documents in strategic writing, there's not just one correct way to prepare a promotional news release. Some include contact information, just like traditional news releases; some don't. Some include symbols such as ©, ® and ™ for the organization's products, although most don't. Some use the pronoun *we* to refer to their own organization, but that approach restricts a journalist's or blogger's ability to republish the promotional news release verbatim: *We* refers to your organization, not to the publication that would like to republish your news release. One of the best ways to learn to write promotional news releases is to study successful examples, such as those within the digital newsrooms of Marvel Entertainment, Spotify and Andrews McMeel Universal.

Many promotional news releases contain four to eight short paragraphs with several links to relevant information and at least one strong image, which sometimes is a video. Promotional news releases often—but not always—contain the following elements:

1. For email promotional news releases, an attention-grabbing subject line that mentions the product—whether that's a person, an event, an object such as a video game or something else
2. An attention-grabbing headline that includes the name of the featured product
3. An optional subheadline, sometimes with an exclamation point. Some promotional news releases feature a straightforward headline and begin the promotional language in a subheadline.
4. A large, attention-grabbing high-resolution image under the headline (and under the subheadline, if there is one). This image generally does not include a caption. Some organizations begin with the image, putting it above the headline.
5. A byline, sometimes with a link to additional information about the author
6. The date and, sometimes, the time (as in 3:15 p.m.) of the posting
7. An attention-grabbing lead paragraph. (For suggestions on writing compelling leads, see page 151.)
8. Details about the product, often with links to additional information
9. A quotation or quotations, often with attributions in present tense (*says*) instead of past tense (*said*). Quotations, however, are not as important in promotional news releases as they are in traditional announcement news releases. A promotional news release is considered to be the voice of the organization. That means you usually don't need to use quotations or paraphrases to attribute important information to reliable sources.
10. An additional image or images, including videos. These images sometimes include captions. For advice on writing captions, see page 134.
11. A closing call to action, such as "Get your tickets now!"

12. A boilerplate paragraph that describes your organization, sometimes with lavish, laudatory language
13. Social media icons that allow viewers to share the release on social media such as Facebook and Twitter

In terms of organization, promotional news releases do not follow an exact inverted pyramid (page 100), in which the most important information comes first and the least important information comes last. Instead, the text of a promotional news release begins with an attention-grabbing lead, which may or may not include the story's most important fact. After that, information does come in order of importance. Unlike an inverted pyramid story, however, the story's conclusion is not the least important section, which could be cut to save space. Instead, the conclusion of a promotional news release often is a call to action or a closing joke, memorable quotation or something equally dramatic. For tips on strong conclusions, see pages 151–152. Again, some promotional news releases end with a boilerplate paragraph that describes your organization.

Promotional news releases generally lack the ### symbol that marks the end of the story in traditional announcement news releases.

## PROMOTIONAL NEWS RELEASE TIPS

1. **Be excited**: Don't overdo it, but do use exclamation points—sometimes even in the headline or subheadline. Marvel Entertainment's promotional news releases usually add an exclamation point to the subheadline. The tone can resemble the tone of a sales message (pages 265–272).
2. **Be subjective**: Go ahead and call your organization "a leader" or "the leader" in whatever business it's in. Promotional news releases can contain unattributed, biased opinions. For example, if you call your new product "revolutionary," do you need a source? Nope—although a trustworthy source might strengthen your point. Don't go overboard, but it's OK to be an entertaining, informative cheerleader in a promotional news release.
3. **Don't go too far**: You can exaggerate as long as reasonable readers will realize you're stretching the truth for entertainment purposes. For example, you can boast that your new racing bicycle is faster than a rocket-sled fueled by concentrated caffeine. Reasonable readers will recognize the technique known as hyperbole. But claiming that your new bike is 17% faster than competing models is different. That's a believable claim, and you shouldn't make it without solid proof.
4. **Know your medium**: Review Writing for the Web, pages 19–25. That section will remind you to consider such things as search engine optimization and word count in the creation of promotional news releases.

## Promotional News Release

Published May 29, 2023 • 4:10 PM

**Starklight Series Pushes Limits with The Dulwich Quest**

MGS Secretly Tested Its Wildest Game Yet!

BY JAMES PIBBLE

Gamers, your wish has been granted! MGS Interactive Games has announced a title and a release date for the new game in its renowned Starklight series. The title of the much-anticipated new entry in the award-winning series will be The Dulwich Quest.

And the news gets even better: MGS recently asked 16 lucky gamers to test drive The Dulwich Quest at four undisclosed locations. MGS president Bert Mulliner says that the gamers raved about the new Starklight entry and that all systems are go for a Sept. 14 sales date.

"We should have waited to announce this, but I'm sick of all the secrecy," Mulliner says. "We're getting upward of 300 inquiries a day, all asking what the name will be and if we've done any test tournaments. Starklight fans are so loyal that it just seemed cruel not to share what we know."

The Dulwich Quest features groundbreaking technology in video games, including multiple team and individual competitions and in-game randomizing that brings new dangers to backtracking.

Can't wait until Sept. 14? MGS feels ya: Click for Dulwich Quest updates and advance purchase!

The Starklight series and The Dulwich Quest are products of MGS Interactive Games, a leading multimedia entertainment company based in Austin, Texas.

SUBSCRIBE / UNSUBSCRIBE

Credit: Ivanko80/shutterstock.com

# 2K Media Advisories

## Purpose, Audience and Media

A media advisory is a *what, who, when, where, why, how* outline of a news story. It is appropriate in two situations:

1. The news is extremely timely—so-called breaking news—and you must get the information to the news media as quickly as possible. You don't have time to write a news release.
2. You want to remind journalists of an important, previously sent news release and to provide basic information that aids coverage of a topic or event.

Journalists are the audience for media advisories. Like news releases, media advisories are sent to journalists in the hope that they will pass along the information to their readers, listeners and viewers.

Media advisories are emailed, tweeted, texted and posted on the homepage of digital newsrooms. Do not text a reporter unless they have agreed to receive your texts.

**KEY TO SUCCESS**

Media advisories should outline only very timely news. A journalist should be able to write a short, complete news story from the media advisory alone—or the media advisory should persuade the journalist to attend a newsworthy event.

## Format/Design

The headings and contact information of a media advisory are the same as those of an email news release (pages 96–97), except that the document is labeled "Media Advisory" instead of "News Release." That distinction is important because a media advisory conveys a sense of urgency.

Media advisories are short and to the point. They are not meant for publication in their present format. They are single-spaced, with double-spacing between paragraphs.

After the headings and headline, most media advisories arrange their information something like this:

**What**: Gov. Jane Smith will tour the Midtown Recycling Center. The tour will be private, but reporters may join. After the tour, the governor will accept questions regarding her visit and her recycling policies.

**Who**: Mike Jones, Midtown Recycling Center founder and president, will conduct the tour for the governor. Midtown Mayor Lynn Johnson will join them. Gov. Jane Smith, an independent, was elected in 2018.

**When**: Saturday, Sept. 14, 3-4 p.m.

**Where**: Midtown Recycling Center, 3309 Riverview, Midtown

**Why**: "I'm visiting Midtown Recycling Center because it's a model facility for the rest of the state. It's the perfect example of my administration's recycling policies."—Gov. Jane Smith

Note how similar the media advisory's format is to the fact sheet's format (pages 139–141). Despite the similarity, fact sheets and media advisories are used for different purposes. A fact sheet accompanies a news release in a media kit or delivers useful facts about people, products or organizations. A media advisory is usually used for breaking stories that don't allow time for the writing of a full news release.

## Content and Organization

The traditional media advisory begins with a traditional news headline (page 99). Following the headline, the advisory becomes a *what, who, when, where, why, how* outline of a story's or event's essential facts. That outline begins with the most important set of facts (often the *what*), then moves to the second most important set of facts (often the *who*) and so on. There's no attempt at a story form. However, either the media advisory should persuade a journalist to attend an event, or the media advisory should be so complete that a journalist could write a short news story from the media advisory alone.

### MEDIA ADVISORY TIPS

1. **Don't overuse**: Use media advisories for breaking news stories or reminders of important, upcoming events, particularly those that may involve potential complications such as media-credentialing processes. Don't issue media advisories for routine stories that can be successfully addressed with a news release.
2. **Follow up**: Because media advisories are comparatively rare and highly newsworthy, you may contact reporters to ensure that they received the advisory and to offer additional help. Avoid this procedure with standard news releases. You can use email and telephone voicemail to follow up.

# Media Advisory (Email)

**To:** JaneQReporter@newspaper.com

**Subject:** Media Advisory: MGS news conference

## MGS Media Advisory

For Immediate Release
Oct. 3, 2023

**MGS Interactive Games schedules news conference to announce Hollywood cross-promotion**

**What:** Bert Mulliner, president of MGS Interactive Games, will announce that company's first video game based on a Hollywood movie at 9 a.m., Friday, Oct. 4, in the MediaDome at MGS headquarters. Mulliner will identify the award-winning movie at the news conference.

**Who:** The movie's award-winning director and top-billed star will make statements, as will Mulliner. All three will take questions. Copies of the statements will be available for members of the news media. Multimedia information on the cross-promotion, including biographies of the director and top star, will be distributed to members of the news media at the 9 a.m. news conference. Afterward, information will be posted on MGS's website.

**Where:** MGS headquarters are at 2010 Ridglea Dr., Austin, Texas. The MediaDome (Room 301) includes lighting suitable for television cameras.

**When:** The news conference will begin at 9 a.m., Friday, Oct. 4, and will last 90 minutes.

**How:** Access to the MediaDome will be through a media credentials checkpoint in the headquarters' third-floor lobby.

###

For More Information:
Jane Doe
Director of Media Relations
555-654-2986
jdoe@mgsintgames.com

# 2L Pitches

## Purpose, Audience and Media

A pitch is a strategic message that attempts to persuade an individual journalist or blogger to write the story described in the pitch. A pitch—whether an email, phone call or social media message through a website such as LinkedIn—promotes, or "pitches," a story idea.

Strategic writers often send a pitch instead of a feature news release (pages 107–109). A pitch promotes a "soft-news" story idea rather than a hard-news story such as the announcement of a new chief executive officer. Don't use a pitch to replace an announcement news release (pages 103–106) that delivers hard news.

Before deciding to send a pitch, consider these facts:

- Written pitches generally take less time to read than feature news releases. Busy recipients will appreciate that.
- In a pitch, as opposed to a feature news release, the story isn't yet written—so a journalist or blogger can feel a stronger sense of ownership of the potential story.
- Unlike news releases, a pitch generally is an exclusive offer sent to one particular journalist or blogger. If the story has relevance for several geographic areas, you can offer it as an exclusive in each area; that is, you can pitch the story to one recipient in each region. However, if the story has national significance, you should approach only one journalist or blogger. If the story has international significance, you could pitch it to one recipient in each nation.

The audience of a pitch is one individual. Ideally, you choose a particular journalist or blogger because you are familiar with their work and know that they'll do a good job on the story. Don't assume, however, that the journalist or blogger will write the story. The goal of your pitch is to persuade them to do so. Research shows that the top four things journalists want to see in a pitch are clear relevance to the journalist's interests and publishing history, an understanding of the journalist's medium, compelling information and assured connections to key sources.

A recent survey by the Go Fish Digital marketing group showed that journalists and bloggers strongly prefer to receive pitches via email, as opposed to phone calls, texts or social media such as Twitter or LinkedIn.

If you know a journalist or blogger well enough to pitch a story via a phone call, don't call during times of day when you know they face a deadline. Avoid phoning on heavy news days when common sense suggests that the writer may be particularly busy.

**KEY TO SUCCESS**

The subject line and first sentences of an email pitch must hook the journalist or blogger. Those opening sections should clearly relate to an interesting, exclusive story.

## Format/Design

For email pitches, follow the email guidelines on pages 96–97.

Pitches increasingly include multimedia elements to show that the story itself could include multimedia features. Consider including a photo and, perhaps, links to other multimedia materials or websites.

## Content and Organization

The subject line of an email pitch must cut through the clutter of countless other emails received daily by your targeted journalist or blogger: Journalists say that only 25% of the pitches they receive are relevant to their personal news beats, so an effective subject line must show the recipient that the story idea fits the subject that they usually cover. The subject line must be clear and concise. Some public relations practitioners recommend calling the journalist by name in the subject line and including the word *Exclusive*—quite a challenge when subject lines must be concise. Such a subject line might read something like: *For J. Hardesty: Exclusive on Top Game Designer*. Such a subject line clearly shows the recipient that the email message is not spam.

Begin your email message with a salutation, such as *Ms. Hardesty*, followed by a colon. (Addressing the recipient by name again shows that this is an individual message, not spam.)

Written pitches, like bad-news correspondence and sales correspondence, are unusual in that they don't use the first paragraph to tell the recipient the main point of the message.

### *Part 1 of 4: Start a Fascinating Story*

Hook the journalist or blogger with the first paragraph. Write the first part of your email as if it were the lead of the story that you hope the recipient will write. Journalists in particular are storytellers, so spark the recipient's attention by beginning an irresistible story.

This first part often will not mention your company or organization. That's because you want to direct the recipient's attention toward the story, not toward promoting your organization. Be concise but specific; journalists and bloggers love details. However, don't hesitate to tantalize the recipient by creating a brief sense of mystery. Make the recipient want to know the rest of the story. Avoid the overused "Did you know . . ." opening.

This first section generally is one brief paragraph. Don't make the recipient wait too long to discover the reason for the email. Don't wait too long to make the pitch.

## *Part 2 of 4: State the Purpose of the Pitch*

Tell the recipient exactly why you're writing—for example, "I think *California Business Today* should do a story on David Smith." (Smith, of course, was introduced in the first section.) If possible, praise a previous story that the recipient wrote; that story, you can explain, is why you think they're the perfect writer for a story on David Smith. (Or, if the pitch is addressed to an editor, mention a recent story in that editor's newspaper section, magazine, newscast or website that was well done.) If you follow the recipient on social media, you increase your chances of knowing what interests them. Give a little more information on David Smith. Continue to tell the story in this section. This section usually is one paragraph.

## *Part 3 of 4: State the Terms of Your Offer*

State that you're offering this idea exclusively to the recipient; that should help gain their interest. Then explain that, because of the exclusive offer, you'll need a reply by a specific date. In the same paragraph, offer help. List the best contacts from your organization and their phone numbers and email addresses (be sure those contacts know you have included them). Offer to help set up interviews and mention multimedia possibilities. Be diplomatic. Now that the recipient knows the idea, they can try to do the story without you—and, of course, they don't have to do the story at all. Don't provoke anger by suggesting that the story can be done only on your terms. Present yourself as a helpful assistant.

This section usually is one paragraph.

## *Part 4 of 4: Describe What You Hope Will Happen Next*

Mention that you'll write again in a few days (name a day, if possible) to see if they are interested and to determine what help you might offer. If your pitch offers a truly good story idea, the recipient often will contact you before your deadline. End with a thank you for the recipient's time and consideration.

## *Follow-Up Telephone Calls*

Survey research shows that journalists and bloggers will accept one follow-up call about a well-written, appropriate, exclusive pitch. (Recall that journalists and bloggers do *not* want follow-up calls about standard news releases; see page 101.) Your follow-up call will often reach only voicemail. If so,

1. State your name and why you're calling: This is Sharon Jones, and I'm following up on Wednesday's exclusive pitch about the new dinosaur fossils in Oregon.
2. Concisely remind reporter of the top benefits/most compelling facts.
3. Remind reporter that this is an exclusive offer—but that you need a response by the specified deadline.
4. Repeat your name. Clearly state—and then repeat—your phone number.
5. Close with a thank you and the hope that you'll hear back soon.

If the journalist or blogger does answer the phone,

a. State your name and why you're calling: *Ms. Manning, this is Sharon Jones, and I'm following up on Wednesday's exclusive pitch about the new dinosaur fossils in Oregon.*
b. Then ask if journalist or blogger has one minute for you. If so, ask if they've had time to consider the story idea and if they have any questions.
c. If reporter resists, politely repeat that this is an exclusive offer.
d. If reporter still resists, thank them and politely conclude the call.

## *Telephone Pitches*

As noted earlier, some journalists and bloggers will consider newsworthy story ideas over the phone, but surveys show that most strongly oppose phone pitches (the recent Go Fish Digital study found that only 2% of journalists prefer phone pitches). Telephone pitches work best when the recipient and the public relations practitioner know and trust each other. Some public relations practitioners, however, are willing to fight the odds and phone journalists or bloggers they don't know. Such calls are known as cold calls.

Like email pitches, telephone pitches generally are exclusive offers. Successful phone pitches share a basic strategy:

1. Thoroughly understand the needs of the recipient and their particular medium (a radio show, for example). Be familiar with the medium's recent content and target audience.
2. Know the daily deadline times of the journalist or blogger you plan to phone—and don't call at those times.
3. Come right to the point. Tell the journalist or blogger you're offering an exclusive story. Concisely explain why the story will interest the recipient's audience.

4. Be well-informed and ready to answer questions about the story.
5. Accept *no* for an answer. Don't damage your relationship with the journalist or blogger by insisting or by asking why they aren't interested.

## PITCH TIPS

1. **Proofread**: Don't rely on your email program's spellchecker. Be sure to proofread the pitch yourself, and have others proofread it.
2. **Sell the story**: Don't focus on how this will benefit your client or organization. The pitch's recipient probably has no interest in promoting your cause or your products, but they do want a good story. Keep the focus on information that will appeal to the recipient and their audience.
3. **Play fair**: Journalists and bloggers will never forget it if you hide bad news or if the story is old news already covered elsewhere. Never waste a busy journalist's or blogger's time. The story you pitch must truly be a good story for the recipient's audience.
4. **Be easy to find**: Include your 24/7 direct phone number in the pitch.
5. **Time the pitch**: The Go Fish Digital survey found that early morning is the best time to email a pitch, and Mondays and Tuesdays are the best days to send a pitch.
6. **Be prepared**: Be ready to respond quickly if the journalist or blogger calls. Don't make them wait for interviews, photographs or any other needs.

# Pitch (Email)

**To:** JaneQHardesty@CalBusinessToday.com

**Subject:** For J. Hardesty: Exclusive on Top Game Designer

Ms. Hardesty:

David Smith hates moonlit nights. In the shadows of his bedroom, the ideas come too quickly—and some are so terrifying that they pin him to the bed. He tries to reach for his phone to record a voice memo, but his hand trembles. "Will this make sense in the morning?" he wonders, torn between hope and fear.

I think the story of David Smith would be ideal for *California Business Today* magazine. David is an award-winning, California-based game designer for MGS Interactive Games. Your recent series on California's creative geniuses has been consistently excellent, particularly your recent story on playwright Cheryl Turcot. As the creative force behind such bestselling games as Night Terror, Are You Sleeping? and Starklight #1, David Smith would be a natural for your creative geniuses series. His best ideas come in dreams so frightening that I sometimes don't want to hear about them. You should see the ideas we reject (unless you're prone to nightmares, we could discuss those).

David Smith

We're offering David's story exclusively to *California Business Today*, so I would need to know by Sept. 28 if you're interested. I can assure you that David would cooperate fully with any writers or photographers you would assign. We also could help you arrange interviews with members of David's staff or with anyone else who works for MGS Interactive Games. We do have photos and videos for you to consider—and we even have recordings of some of David's midnight voice memos. (Yes, his voice does tremble in those.)

Please let me know how I might help. My direct phone line is 555-498-8871.

Thank you for your time and consideration.

Angie Perez

Angie Perez
Public Relations Director
MGS Interactive Games
555-498-8871
ap@mgsintergames.com

Credit: lassedesignen/shutterstock.com

# 2M Video News Releases and Direct-to-Audience Videos

## Purpose, Audience and Media

Contrary to what many may believe, charges of so-called fake news are not really new. In fact, that phrase was the headline on the cover of *TV Guide* magazine in February 1992. Whatever connotation it may have today, the magazine article referred to the use of video news releases (VNRs) within newscasts. This practice was—and still is—considered controversial because television and web-based journalists did not always distinguish between their own original content and that produced by outside sources.

First introduced in the 1970s, VNRs are video versions of news releases. Much like the traditional news release, a VNR provides ready-to-use content to journalists with the goal of having them share it with their audiences. When properly used, VNRs are especially effective during crises, making announcements or providing journalists with video content that might otherwise be unattainable, such as scenes from a restricted-access laboratory or a remote location. However, when either the creator of the VNR or the news organization using it fails to properly identify its source to the public, it then becomes controversial and may run afoul of legal and/or ethical standards.

Like all media, VNRs have evolved in the Digital Age. They are now used on a variety of platforms—many of which *TV Guide* could not envision in 1992. VNRs are no longer just for television journalists: Videos have become an essential part of online news sources. And the advent of social media has made it possible to bypass journalists and reach out directly to people through direct-to-audience videos. In this environment, VNRs have become helpful tools in allowing organizations to tell their story to a targeted audience.

"It's not news to anyone when I say we've entered a time where communications have become ubiquitous, and we live amongst a generation of hyper-connected individuals," said Alan Berkson of Freshdesk, a cloud-based customer service platform. In an interview with ZDNet, Berkson said, "With the abundance of information—useful and otherwise—available to today's consumers, it's becoming more difficult to make an impact and be heard above the noise."

In an attempt to be heard above the din, many organizations are engaging in content marketing (pages 245–246), storytelling that delivers messages through compelling narratives. Instead of relying only on a fact-based, feature-benefit approach to persuasion, they are telling stories that not only provide information but also motivate listeners by engaging them with narratives about the real-life experiences of people to whom they can relate. While content marketing involves a variety of media, video—in the form of video news releases (VNRs) and direct-to-audience videos—is an especially effective medium for making an emotional connection.

As noted, Digital Age technology has made it easier for organizations to send unfiltered messages directly to the public without journalist gatekeepers. But there is a downside: Since journalists do not filter the content, direct-to-audience videos do not necessarily have the credibility that comes with third-party endorsement.

**KEY TO SUCCESS**

VNRs and direct-to-audience videos must contain information relevant to the target audience. In the case of VNRs, they must also serve the needs of journalists.

## Format/Design

The role of a strategic writer in the production of a VNR or direct-to-audience video is to produce a two-column script (described in more detail in Video Advertisements, pages 210–216). The Table and Borders function in most word-processing programs allows you to create visible borders around the two columns.

The migration of VNRs to digital newsrooms and the internet has changed their packaging. VNRs that directly target the primary audience through YouTube or other social media appear as self-contained reports/stories that require little explanation. Any essential additional information accompanies the video's link. However, more traditional VNRs accommodate journalists' needs. In this case, the goal is to give journalists editorial flexibility to meet each media outlet's particular needs.

## Content and Organization

Audience-direct VNRs—those that bypass journalists—are often embedded within organization websites, including digital newsrooms, or are featured on organizational YouTube channels. Those websites, in turn, can contain text, pictures and links to a variety of related information. The audience is invited to visit the website or VNR through an email or links embedded within social media. In social media, the video itself, with introductory text, can be the post. In the case of email, the invitation often includes eye-catching graphics.

VNRs targeting television and web-based journalists must be self-contained. By that, we mean the VNR provides a reporter or editor all of the information needed for writing and producing a story. Journalist-targeted VNRs have five parts, sometimes delivered as separate digital files:

1. *Opening billboard.* This provides the essential background information about the news story that follows. It clearly identifies who produced the VNR and includes contact information for follow-ups. It identifies sources interviewed on-camera by name and title. It also includes a suggested announcer lead-in to the story, the running time of the story and its outcue (or closing words).
2. *Video news release.* This is a 60–90-second video news story, with an announcer voiceover, natural sounds (SOT, meaning "sound on tape") and soundbites. Two major differences separate the VNR from the news story you see on television news. First, the "reporter" (who is not really a journalist) is not seen on camera. Second, it is a "rough cut," lacking titles and graphics. The reason for these two differences is the same: Stations want flexibility to add their own people and graphics packages to customize the information to fit an established station image.
3. *Video news release without announcer voiceover.* The only audio in this version of the VNR is the natural sound within the taped footage (SOT) and the soundbites. This gives editors additional flexibility, allowing them to use their own personnel. A brief billboard announcing the deletion of the voiceover precedes this section.
4. *Soundbites.* These include a variety of soundbites, particularly some that were not in the VNR. A billboard of who is speaking, the length of the soundbite and, often, a suggested lead-in precede each soundbite. The more variety, the greater the likelihood that journalists may use the story in multiple newscasts.
5. *B-roll.* This is additional footage journalists can use to illustrate the story. A brief billboard that specifies content and running time precedes this section.

Before writing a VNR or direct-to-audience video script, be sure to review Writing for the Ears, pages 29–34.

## VIDEO NEWS RELEASE TIPS

1. **Think again**: Before investing the time and expense of producing a video, make certain that this tactic fits within your organization's overall strategy. Always ask if the video is the best way to achieve a desired goal. In the case of a VNR, are there legitimate reasons why news organizations would not be able to generate the same footage and interviews on their own? When news editors commit time and resources to covering a story, they are more likely to air it.
2. **Check your ego at the door**: Only rarely do television and web-based news organizations take a VNR and air it without any changes. Just as with a traditional news release, a VNR often is a journalist's starting point.
3. **Engage**: Tell a story that connects with the audience. How does your message relate to the target public? In addition to providing information, try to connect with the audience on an emotional level. You can achieve that connection if you have thoroughly researched the demographics and psychographics of your target.
4. **Remember that news is still news**: Journalists will not use anything overtly promotional. If the message in a VNR is a selling message, news producers and reporters won't use it. The same is true for direct-to-audience videos. Remember: This is the most media-savvy generation in history.
5. **Don't do it on the cheap**: Many organizations have excellent in-house video production facilities. However, most do not. In those cases, consider hiring a competent agency or independent producer. If a VNR has poor production values, it will not be used.
6. **Make pictures and words work together**: Remember the concept of hit-and-run writing, discussed in Writing for the Ears (pages 29–34).

## Video News Release Production Script

F4U FDN - 1

| **Project: Flavours4U Foundation VNR** | |
|---|---|
| CHYRON–PAGE ONE: Cardiovascular disease (CVD) claims the lives of nearly 1 million Americans every year. CVD is the leading cause of death in the United States. According to the American Heart Association, at least 58 million Americans—one out of every four—suffer from some form of heart disease.<br><br>CHYRON–PAGE TWO: In an effort to reverse this trend, CEO Michelle Campion of Flavours4U Products Inc., the maker of a popular pre-packaged diet program, announces the creation of the Flavours4U Foundation for the purpose of supporting CVD research. To launch this research, Flavours4U Products Inc. is contributing $1 million in seed money. The company also hopes to raise an additional $4 million through a Have a Healthy Heart Campaign.<br><br>CHYRON–PAGE THREE: This is a video news release on the Have a Healthy Heart campaign. It will feature interviews with Dr. Marco Podestra (poh-DESS-trah), chief cardiologist at Riverview Community Medical Center and Paula Pressman, executive director of the Flavours4U Foundation. The announcer is Robert Franks. For more information, contact Flavours4U Foundation Executive Director Paula Pressman at 999-555-1040, ppress@F4Uinc.com or log on to www.F4Uhealthyheart.com.<br><br>CHYRON–PAGE FOUR: Suggested lead-in: "When it comes to fighting heart disease, the creators of a popular diet plan want to put their money where your mouth is . . ."<br><br>RT: 1:00<br><br>FIVE SECONDS OF BLACK | |

# Video News Release Production Script

F4U FDN - 2

| | |
|---|---|
| OPENING SEQUENCE OF SHOTS—Dr. Marco Podestra giving a checkup to a heart patient. | ANNOUNCER: Approximately one-million people in the United States will die from cardiovascular disease this year. And the American Heart Association says one out of every four Americans has some form of heart disease. That's more than 58-million people. Cardiologist Marco Podestra of the Riverview Community Medical Center says this is an alarming epidemic. |
| CU: Dr. Podestra in lab. | DR. MARCO PODESTRA: It is unbelievable. In a country where the world's best health care is at our fingertips, I don't understand how one out of every four Americans faces some form of heart disease. We have to do something about that. |
| SEQUENCE: Exterior shot of Flavours4U Products Inc., followed by shots of products. Includes close-up of special markings on packaging. | ANNOUNCER: In an effort to stop the nation's number-one killer, Flavours-4-U Products is launching the Have a Healthy Heart Campaign. A portion of the proceeds from specially marked *flavours!* diet products will be placed in a fund created to support heart research. Paula Pressman is |

## Video News Release Production Script

F4U FDN - 3

| | |
|---|---|
| | the executive director of the Flavours-4-U Foundation. She says it is in everyone's best interests to get involved. |
| CU: Paula Pressman in Flavours4U Foundation offices with Have a Healthy Heart logo clearly visible in the background. | PAULA PRESSMAN:<br>It is not enough for our company to make products that reduce the risk of heart disease. We want to join with the dedicated doctors and researchers who are searching for ways to prevent cardiovascular disease. You might say our heart is into this cause. |
| CLOSING SEQUENCE—Heart researchers working with patients in a medical laboratory setting. Closing shot is a close-up of a working heart monitor. | ANNOUNCER:<br>To get the Have a Healthy Heart Campaign off to a good start, Flavours-4-U Products has donated one-million dollars to the effort. Pressman says the company hopes to raise another four-million dollars in the fight against heart disease. I'm Robert Franks reporting. |
| FIVE SECONDS OF BLACK | |
| CHYRON: Soundbite #1—Chief Cardiologist Marco Podestra of Riverview Community Medical Center. RT:13. | PODESTRA:<br>Of the 50-million Americans who have high blood pressure, one-third of them don't know they have it. Every 29 seconds, someone in |

# Video News Release Production Script

F4U FDN - 4

| | |
|---|---|
| | the U-S suffers a coronary event. About every 60 seconds, someone dies from one. |
| FIVE SECONDS OF BLACK | |
| CHYRON: Soundbite #2—Paula Pressman, executive director, Flavours4U Foundation. RT:14 | PRESSMAN:<br>We are hoping to raise an additional four-million dollars to support heart research. And that is on top of the one-million dollars Flavours-4-U Products has already dedicated to this effort. Heart disease touches one in four Americans. We want to help reduce the threat. |
| FIVE SECONDS OF BLACK | |
| Roll with natural SOT; RT 1:30 B-ROLL: A series of shots of Dr. Podestra examining patients. Heart research in medical laboratory. Pressman and foundation workers planning the campaign. Products with special markings. | |

###

# 2N Digital Newsrooms and Media Kits

## Purpose, Audience and Media

Digital newsrooms and media kits are storehouses of information for journalists, bloggers and others seeking information. A digital newsroom is an organization's web-based repository of new releases, reports, videos, fact sheets, biographies and histories, links to media coverage of the organization, information for investors, contact information and more. One of the best ways to understand digital newsrooms is to study good examples. Digital newsrooms that have won praise from journalists and public relations professionals include those of Allstate Insurance, NASA, the American Lung Association and Ketchum.

A media kit is, in a sense, an expanded news release. Media kits contain at least one news release. They also contain other documents, such as backgrounders and fact sheets (pages 135–141). Media kits also can contain photo opportunity advisories, captioned photographs, business cards, product samples and other items that help tell a story. The purpose of a media kit is to deliver more information than a news release alone could supply. For example, television networks often prepare media kits about upcoming television seasons.

Media kits often are called press kits, a term that no longer seems appropriate. Today, of course, we get news not only from print news media (which use a printing press) but also from television, radio, the web and social media. The term *media kit* is more accurate than *press kit*. However, the term *media kit* also describes a packet of information that a medium such as a magazine prepares for advertisers. That type of media kit includes advertising prices and reader profiles, among other items.

The primary audiences for digital newsrooms and public relations media kits are journalists and bloggers. Just as we said for news releases, journalists have made clear what kinds of information they want from organizations: newsworthy, timely, unbiased facts and multimedia materials. However, an information-packed digital newsroom can be a great source for anyone seeking information about a particular organization.

The medium of digital newsrooms is the web, making them ideal for the storage and distribution of vast amounts of information and multimedia materials.

Like news releases, media kits exist in a variety of media. Modern email news releases actually function as media kits by embedding links to additional information: For example, clicking on a highlighted company name in a news release would lead a journalist to a biography of that company, which itself could link to a digital newsroom that includes a video biography. Some media kits are still produced on paper and packaged in attractive folders; such kits generally are for distribution at trade shows. Even at trade shows, however, USB flash drives are replacing old-fashioned paper media kits.

**KEY TO SUCCESS**

Digital newsrooms and media kits supply newsworthy, timely information with significant amounts of background documentation, including multimedia materials.

## Format/Design

Digital newsrooms are well-organized with highly visible navigation menus that lead visitors to such categories as news releases and videos (see Websites, pages 84–93). Within each category, a table of contents lists document titles/links, and a photo or other image often accompanies each document title. For example, one section of a digital newsroom could list news releases by title, starting with the most recent. Each title would be a link and would be accompanied by a photo or other image.

Media kits exist in a variety of formats. Any format that gathers and organizes a small number of documents can be suitable for a media kit. As noted earlier, email news releases with embedded links can serve as media kits. For trade shows, some paper media kits are packaged in folders with internal pockets, just like those students sometimes use for classes (though the folders often feature an organization's logo). Some paper media kits appear as small boxes with folders inside. Those folders include news releases, backgrounders, fact sheets and other documents. And, again, some media kits exist as documents stored on a USB flash drive.

## Content and Organization

Digital newsrooms are highly compartmentalized, containing categories such as organization overview, news releases, photographs and images, videos, blogs, profiles of organization leaders, media coverage of the organization, links to the organization's social media and more. Companies that sell stock sometimes have separate investor relations sections with financial news.

A media kit must contain at least one news release (pages 94–102). Other traditional documents in a media kit include backgrounders, fact sheets and photo opportunity advisories. Those documents are described in the sections that follow.

## NEWSROOM/MEDIA KIT TIPS

1. **Monitor**: Constantly review your organization's digital newsroom for flawed links. Journalists on deadline need quick access to accurate information.
2. **Compare**: Examine other organizations' digital newsrooms. Publications such as *PR News* have annual competitions for public relations tactics, including best digital newsrooms.
3. **Include samples**: Journalists and bloggers are human. In trade show paper media kits, they like toys, gifts, novelties and free samples. When appropriate, include a product sample or a novelty with your organization's name on it. Avoid expensive gifts that might appear to be bribes.
4. **Write captions for photographs**: Study traditional media sources for guidance on writing good photo captions. The first sentence of a caption generally is in present tense. That first sentence acknowledges the scene in the photograph, but it also tells the reader something more. For example, "Acme Widget employees applaud as the millionth widget rolls off the assembly line." The photo shows people applauding and the rolling widget, but the caption adds why *this* widget is important.

   Because the first sentence is in present tense, it shouldn't include a *when*, which would tear the sentence between present and past tense. Put the *when* in a second, past-tense sentence—for example, "Acme Widget employees applaud as the millionth widget rolls off the assembly line. Workers in the Park City factory paused Tuesday, Oct. 1, to celebrate the event." Captions rarely exceed two sentences.

   Some captions begin with a boldface teaser sentence fragment followed by a colon—for example, "**A World of Widgets**: Acme Widget employees applaud as the millionth widget rolls off the assembly line." Instead of the term *caption*, journalists often use the term *cutline*.
5. **Lead with the news**: In old-fashioned paper-folder media kits, place the news release on the right-hand side of the folder. If you have a fact sheet, place it on the left-hand side. Place any backgrounders behind those two documents. However, if you have an important photo opportunity advisory, place it on the left instead of the fact sheet.

# 20 Backgrounders

## Purpose, Audience and Media

A backgrounder supplements a news release. For example, a backgrounder can be a biography of a key individual mentioned in the news release. Another backgrounder for the same news release could be a history of the relevant organization. News releases do not rely on backgrounders to deliver the essential information in a news story; a news release should be complete within itself. However, journalists and bloggers appreciate details and extra, relevant information—and backgrounders are perfect for delivering such extras. A backgrounder can be a link in an email news release, a document in an organization's digital newsroom or "About Us" website section or a document in a paper media kit.

The audience for a backgrounder when it exists as a link in a news release is a journalist or blogger. When a backgrounder exists as a document or page in an organization's digital newsroom or "About Us" section, its audience is anyone seeking information about the organization.

**KEY TO SUCCESS**

Backgrounders are not news releases. They do not have news leads or news headlines. They supply interesting, relevant background information. A news release must be able to stand alone without a supporting backgrounder.

## Format/Design

The headings for digital and PDF/paper backgrounders are like those of email and PDF/paper news releases (pages 96–97) except that instead of "News Release" in large type, a backgrounder, of course, has "Backgrounder."

Give your backgrounder a headline, which usually is simply the title of the subject matter of the backgrounder—often just the name of a person or of an organization.

As with news releases, if a PDF/paper backgrounder is more than one page, type "-more-" or "-over-" at the bottom of each appropriate page. Beginning with the second page, place a condensed version of the backgrounder's headline (called a "slug") and the page number in the upper-right corner. After the last line of the backgrounder, space down two lines and type "-30-" or "###."

Staple the pages of a paper backgrounder together. Don't use a paper clip.

Like news releases, backgrounders are single-spaced and appear in a ready-to-publish format. Like news releases, backgrounders should rarely exceed the equivalent of two paper pages.

## Content and Organization

Unlike news releases, backgrounders are not important news stories. Therefore, backgrounders do not have newspaper-style headlines. Instead, a backgrounder headline generally is the name of the person, organization or event being described.

Backgrounders should not have a *who, what, where, when, why, how* lead—that's the job of news releases. Backgrounders aren't newsworthy. They don't have datelines in the lead. Backgrounders supply background information to news stories. Backgrounders are much more like encyclopedia entries than news stories.

For example, if you look up Abraham Lincoln in an online encyclopedia, you won't find a news flash about his assassination. Instead, you'll find a simple headline (his name) and a beginning sentence that identifies him as the 16th president of the United States. After a very concise description of the reasons for his fame, the entry would begin at the beginning and move in chronological order. Probably, the entry would begin like this: "Abraham Lincoln was the 16th president of the United States and the first U.S. president to be assassinated. Born in a log cabin in Kentucky in 1809, he. . . ."

A familiar way to think of a backgrounder is to compare it to a link. If you were reading an online news story and saw a particular name highlighted in blue, you would know that clicking on that name would link you to that individual's biography. Imagine that the main news story is the news release and the linked biography is the backgrounder.

The standard organization of a backgrounder begins with a defining sentence, then moves to a description of the beginning of the person, organization or event, and then moves in chronological order from the past to the present—at which point the backgrounder ends. If the backgrounder describes a current employee, a concise description of job duties often follows the opening defining sentence.

Your backgrounder should not contain information that should be in a news release that it accompanies; the news release must be able to stand by itself. Some overlap of information in a news release and a backgrounder is inevitable. For example, a news release on a corporate award might note that Jane Smith is the CEO. An accompanying backgrounder on Smith also would note that she is CEO, but it would include additional personal and professional information not in the news release.

Like a news release, a backgrounder should not include unattributed opinions. The tone should be objective, just like an encyclopedia entry. Any opinions should be attributed to clear and credible sources.

Backgrounders rarely are published. Journalists may use them to ensure that they understand the news release, or they may pull a paragraph or two from a backgrounder and insert them into the news release.

## Backgrounder (PDF/Paper)

2010 Ridglea Drive • Austin, TX 55111 • 555-999-5555

# Backgrounder

**FOR IMMEDIATE RELEASE**
May 29, 2023

**FOR MORE INFORMATION:**
Ivy Flora
Investor Relations Director
(555) 123-1120
iflora@mgsintgames.com

**MGS Interactive Games**

MGS Interactive Games creates, manufactures and distributes video games. The company has 430 employees and, in 2022, had revenues of $840 million.

MGS was founded in Austin, Texas, in 2006 by Bert Mulliner, the company president. Mulliner said that the M in the company's name stands for his last name but that the G and the S stand for George (Gershwin) and Stevie (Ray Vaughan), two musicians whom Mulliner listens to when he designs games.

The company's first game, distributed in 2007, was Odessa Overhang, which won the 2009 Game of the Year Award from VPlayToday magazine. In 2005, MGS launched the Starklight series, created by award-winning designer David Smith, with the game Pelican Abbey. In September 2023, MGS will release the fifth game in the Starklight Series, The Dulwich Quest. According to VPlayToday magazine, the Starklight series is the second-best-selling video game series in history, trailing only Quaxart's Jupiter937 series.

MGS has created 23 different video games since its founding.

The MGS Statement of Values identifies creativity, collegiality and community as the three ideas that unite the company. To honor the value of community, the company donates 5% of annual revenues to community-development programs in Austin. Mulliner said that MGS also seeks ethnic diversity among the 20 interns that the company hires annually.

MGS Interactive Games is a publicly held company trading on the New York Stock Exchange under the symbol MGSI.

###

# 2P Fact Sheets

## Purpose, Audience and Media

The traditional fact sheet is a *what, who, when, where, why, how* outline of the accompanying news release in a media kit. Some editors prefer this stripped-down presentation to the news release. Broadcast editors may prefer fact sheets because they're not written as newspaper stories. Other editors may prefer fact sheets because, by listing only facts, they avoid the subjectivity that many editors fear lurks in news releases. Fact sheets generally appear only in media kits (pages 132–134).

The audience for a fact sheet is journalists and bloggers. Fact sheets exist as paper documents in traditional paper media kits, such as those distributed at trade shows, and as documents in large digital media kits, such as a kit that announces the programming for a network's new television season. In such a kit, each show might have its own fact sheet in addition to a show-related news release. In such digital media kits, facts sheets often exist as PDFs.

**KEY TO SUCCESS**

A fact sheet must be so complete that a journalist could write a short news story—often called a news brief—using only the fact sheet.

## Format/Design

The headings for a PDF/paper fact sheet are like those of a news release (pages 94–102) except that instead of "News Release" in large type, a fact sheet, of course, has "Fact Sheet." Fact sheets rarely exist in an email format.

Give your fact sheet the same newspaper-style headline you used for the related news release.

Fact sheets are single-spaced, with double-spacing between paragraphs.

Do your best to keep a fact sheet to one page. After the last line of the fact sheet, space down two lines and type "-30-" or "###."

## Content and Organization

After the headline, a fact sheet begins with the most important information (almost always the *what*) and then moves to the second most important information (almost always the *who*). From there, the general order is *when* and *where*—and, if necessary, *why* and then *how*. This order can change, depending on the importance of the information in each section. The more important the information, the higher it should be on the page.

As the name suggests, a fact sheet covers just the facts. There's no attempt at a story form. However, the fact sheet should be so complete that a journalist could write a short news story based on the fact sheet alone.

Everything in the fact sheet should be in the news release. However, the reverse is not true. Not everything in the news release need be in the fact sheet—just the essential details. For example, a news release may contain quotations that do not appear in the fact sheet.

Most fact sheets arrange the text below the headline in this way:

**What:** The XYZ Corp.'s annual barbecue for the United Way. Beef, pork and vegetables will be grilled.

**Who:**
- The XYZ Corp. is the largest employer in Central City. It makes shoelaces for . . .
- All residents of Central City are invited.

**When:** Sept. 14, 5–9 p.m.

**Where:** Central City Park, 193 Main St.

**Why:** The barbecue annually raises more than $10,000 for local charities.

**How:** Donations of $5 per attendee will be accepted at the park gates. Registration is not required. XYZ Corp. will supply all food and eating utensils. Attendees should bring blankets or lawn chairs.

A different category of fact sheet functions as a backgrounder. This kind of fact sheet does not summarize the accompanying news release. Instead, it provides interesting background facts that support the story in the news release. Unlike a backgrounder, however, this kind of fact sheet is not written as a story. It simply is a list of facts; often, each fact is introduced by a bullet (•). For example, a media kit for a basketball tournament might have fact sheets that list each team's record for the season: wins, losses and related scores for each team. This kind of fact sheet has a backgrounder-style headline rather than a news release–style headline.

## Fact Sheet (PDF/Paper)

2010 Ridglea Drive • Austin, TX 55111 • 555-999-5555

# Fact Sheet

**FOR IMMEDIATE RELEASE**
May 29, 2023

**FOR MORE INFORMATION:**
Catherine Jones
Public Relations Director
(555) 123-4567
cjones@mgsintgames.com

### MGS Names and Tests New Starklight Game

**WHAT**

- The Dulwich Quest is the name of the new game in MGS' Starklight series. MGS Interactive Games of Austin, Texas, produces the Starklight series.
- MGS tested The Dulwich Quest at four undisclosed U.S. locations on May 25.
- Gamers in each location signed pledges of secrecy in return for free copies of the game when it debuts in September.
- According to MGS President Bert Mulliner, the gamers gave The Dulwich Quest "rave reviews." They offered three suggestions for improvement that MGS designers are now testing.
- The Dulwich Quest is the fifth game in the Starklight series, which began in 2005 with Pelican Abbey.
- The Starklight series features games that pit individual players against programmed competitors or live (in-person and online) players.

**WHO**

- The news was announced by MGS President Bert Mulliner. Mulliner founded MGS Interactive Games in Austin, Texas, in 2006.
- MGS has 430 employees and, in 2022, had revenues of $840 million.

**WHERE**

- MGS Interactive Games is in Austin, Texas, at 2010 Ridglea Dr.
- The secret testing of The Dulwich Quest occurred in four undisclosed U.S. cities.

**WHEN**

- Mulliner made the announcement on May 29. The secret testing of The Dulwich Quest occurred on May 25.
- The Dulwich Quest will be available for purchase on Sept. 14. Suggested retail price is $60.

###

# 2Q Photo Opportunity Advisories

## Purpose, Audience and Media

As its name suggests, the photo opportunity advisory is designed to attract photographers and videographers to an event you're publicizing. When you have a newsworthy, visually interesting event that you want to publicize, send photo opportunity advisories to print, TV and online journalists—and even many radio stations have multimedia websites.

Not all events merit a photo opportunity advisory. For example, a news release announcing a corporation's quarterly profits probably lacks a related visual event. But if your organization sponsors a skateboard race between your CEO and the mayor—all for charity—you have a great photo opportunity.

Photo opportunity advisories sometimes are part of media kits (pages 132–134). However, they can be sent individually to journalists and bloggers. Like news releases, photo opportunity advisories can be distributed via email and announced via social media such as Facebook and Twitter. They also can be posted in digital newsrooms.

**KEY TO SUCCESS**

A photo opportunity advisory must move quickly to a detailed, engaging description of a forthcoming visual event.

## Format/Design

The headings for a photo opportunity advisory are like those of email news releases (pages 96–98) except that instead of "News Release" in large type, a photo opportunity advisory, of course, has "Photo Opportunity."

The subject line of an email photo opportunity advisory includes the words "Photo Opportunity" or, to save space, "Photo Op."

Photo opportunity advisories are single-spaced, with double-spacing between paragraphs.

Photo opportunity advisories often open with a descriptive, promotional paragraph. After that paragraph, they adopt the concise format of a fact sheet (pages 139–141), focusing on *what, who, where, when, why* and *how*. After the last line or item, space down two lines and type "-30-" or "###." Close with For More Information details.

## Content and Organization

Unlike most other media kit documents, photo opportunity advisories can have promotional writing. Even the headline of a photo opportunity advisory can be promotional and subjective. Focus the headline on the photogenic nature of the event.

After the headline, write a concise, detailed, promotional paragraph showing why the event is photogenic and newsworthy. Don't go overboard—don't promise more than you can deliver—but journalists will accept promotional writing in this passage.

After the opening paragraph, adopt the style of the fact sheet, focusing on *what, who, where, when, why* and *how*. Be specific about times and places. Some photo opportunity advisories can link to maps to show photographers where the event will be.

In the *how* section, consider specifying what equipment and facilities, including electrical outlets, will be available for photographers.

### PHOTO OP TIPS

1. **Cover it yourself**: Be sure to photograph and film the event yourself. Be ready to supply information and images to journalists who could not attend.
2. **Follow up**: Photo opportunity advisories are distributed *before* an event, of course. After a newsworthy, photogenic event, consider posting a news release with links to your own high-resolution photos and videos of the event.

# Photo Opportunity Advisory

**To:** JohnPhotographer@newspaper.com

**Subject:** Rubber duck race on Patterson Creek

## Photo Opportunity

**For Immediate Release**
**April 23, 2023**

**MGS to sponsor duck race for Donors United**

*Thousands of bright yellow rubber ducks surging down Patterson Creek.... Hundreds of cheering spectators urging them on, all hoping that their entry will win them $100 at the Old Bridge finish line....*

**WHAT**
- MGS Interactive Games will sponsor a rubber duck race on Patterson Creek on Friday, June 7.

**WHY**
- The event is a fundraiser for the Travis County Chapter of Donors United.

**WHERE**
- The race will begin at McDaniel Park and end at Old Bridge.
- MGS Interactive Games is in Austin, Texas, at 2010 Ridglea Dr.

**WHEN**
- Ducks will be launched at noon, Friday, June 7.

**WHO**
- MGS intern Andrea Smith, a Palmquist University sophomore, developed the idea.
- The event will benefit Donors United, a clearinghouse for Travis County charitable organizations.
- Anyone 18 or older may enter.
- MGS Interactive Games creates, manufactures and distributes video games. The company has 430 employees and, in 2022, had revenues of $840 million.

**HOW**
- Ducks cost $5 for two entries and can be purchased at MGS headquarters at 2010 Ridglea Dr. from Monday, May 5, through Thursday, June 6.

###

For More Information:
Jane Doe
Director of Media Relations
555-654-2986
jdoe@mgsintgames.edu

# 2R Newsletter and Magazine Stories

## Purpose, Audience and Media

A newsletter/magazine story is a narrative that delivers facts on an important subject to a large, well-defined audience. A newsletter/magazine story usually can be read in one sitting and is designed either to inform or to entertain and inform. Some newsletters are published daily. However, most newsletters and magazines are weekly or monthly. Therefore, newsletter and magazine stories rarely announce breaking news that readers must know right away.

Most newsletters and magazines are considered niche publications. Niche publications target well-defined audiences whose members share a common interest. Newsletters and magazines can target an organization's employees; members of a profession, such as accountants; members of an association, such as the American Library Association; people with a common interest, such as antique cars; and so on. Newsletters and magazines that target customers and potential customers are an important part of the emerging area of content marketing (pages 245–246).

Newsletters and magazines exist as webpages and paper products. Readership and profits for digital magazines are growing, and even paper magazines have increased profits by decreasing the number of issues per year and increasing issue size. Newsletters are often distributed as email messages with links, and many digital magazines email their tables of contents, with links, to interested recipients.

**KEY TO SUCCESS**

Newsletter and magazine stories must quickly show readers why they'll benefit from reading the story. Readers must quickly realize that they'll learn useful information and that they might even be entertained as they learn.

## Format/Design

Newsletter/magazine stories ultimately will be formatted by a publication designer. In all likelihood, you will submit your stories to an editor via a shared website or as a file attached to an email message. Always consult your editor for format preferences.

At the top of your article, include a proposed title, a subtitle if necessary and a byline (your name).

Single-space your manuscript, unless your editor instructs otherwise.

Use an extra line of space between paragraphs to indicate the beginning of a new paragraph (unless, again, your editor tells you otherwise).

Hit your space bar only once, not twice, after periods and other punctuation marks that end sentences. Have only one space, not two, between sentences.

At the end of your story, type "-30-" or "###" or some other closing symbol.

## Content and Organization

Newsletter stories generally are short and tightly constructed. In most newsletters, space is at a premium. Magazine stories, however, can be longer and can more thoroughly develop a topic.

Most newsletter and magazine stories fall into one of three categories:

1. Straight news stories or announcements
2. Feature stories
3. Hybrid stories

Within these three categories, you can offer an almost endless variety of stories to many different audiences: so-called hacks and how-to instructions, recipes, compilations of consumer-generated content (such as shared advice and/or photos), multimedia stories on topics of interest (including stories about your organization and its employees), entertaining or informative images or infographics, white papers (fact-filled reports), case studies, interviews with interesting people, real-time question-and-answer sessions, news about popular products and much more.

The designer of the newsletter or magazine often will be responsible for finding multimedia material, such as photographs, to accompany your story—but don't hesitate to make suggestions.

### *Straight News Stories*

Straight news stories begin with a headline that summarizes the story's main point. The lead (a first paragraph) of a news story includes the most important details of *who, what, when, where, why* and *how*. The first sentence of the story often includes *what, who, where* and *when*. Leads don't have to include every detail of *who, what, when, where, why* and *how*; that's what the rest of the story is for. Leads should include only the basic, essential details that a reader must know.

In a straight news story, information appears in descending order of importance—the inverted pyramid organization (page 100). The last paragraph is the least important paragraph and could even be deleted from the story without serious damage.

The straight news story is, except for the format, basically identical to the announcement form of the news release (pages 103–106). Straight news stories focus on informing readers, not on entertaining them.

### *Straight News Story Example*

**Palmquist students arrested for property damage**

Eight Palmquist University students were arrested Saturday on charges of damaging grass in front of Rogers Hall during a football game. The misdemeanor charge could result in fines and a jail sentence.

"The students were incredibly irresponsible," said Charles Poole, clerk of the City Municipal Court. "Campus police say that the grass is badly damaged. But I doubt that they'll pay a fine or go to jail if they're guilty. In cases like this, people usually get sentenced to a certain amount of community service."

Palmquist University President Bertrand Bishop said it was important for current students to remember that they are caretakers of the university for future students.

The eight students, all from Central City, are Derek Barrett, 19; Alec Duncan, 20; Rachel Tobin, 18; Jocelyn Snyder, 21; Kendra Jessie, 18; Maggie-Sophia Smith, 20; Gillian Williams, 21; and Emily Davidson, 20.

## *Feature Stories*

Like straight news stories, feature stories inform—but they also entertain. Storytelling skills are important in features. For example, the headline of a feature usually doesn't summarize the story's main point. Instead, it teases and beguiles readers, making them want to read the story to satisfy their curiosity. If necessary, a subheadline can summarize the content of the story.

A feature lead—unlike the lead of a straight news story—need not present all the most important facts. Instead, the first paragraph should "hook" readers, making them want to read the story.

You can present details creatively in a feature. For example, a writer once was assigned a story on the sale of a national retailer's one-millionth pair of socks. Instead of writing a straight news story, he calculated that a kicking chorus line of one million socks wearers would stretch from New York City to Little Rock, Arkansas. He then determined how far into space the unraveled thread from one million pairs of socks would reach. Readers praised and remembered the story; they were informed and entertained.

Standard advice for writing feature stories is *show; don't tell*. For example, don't *tell* readers that Jane Smith is happy; instead, use anecdotes and descriptive writing to *show* that Jane is happy. Showing rather than telling engages readers in the feature; they see the evidence and draw their own conclusions.

In a straight news story, the last paragraph includes the least important information. In a feature, however, the conclusion may be the most important moment of the story; a feature's conclusion may have important dramatic value and provide a sense of closure.

### *Feature Story Example*

**Nobody kicks Palmquist's grass**

Injuries and penalties are nothing new to football, but several Palmquist University students were caught off guard, so to speak, when campus police cited them for unnecessary roughness to grass.

On the defensive are eight Palmquist students who scrimmaged last October on the rain-soaked lawn in front of Rogers Hall. Fifteen yards and a loss of down apparently won't suffice: The undergraduate sodbusters will tackle the local legal system this month, and the misdemeanor charge of tearing up turf carries the possibility of both a fine and a jail sentence.

An official of the City Municipal Court recently said that although the Palmquist Eight had walked all over the rights of the grass, he doubted that they faced a stint in jail. If found guilty, he said, the students may have to pay their debt to society through a specified amount of community service.

Meanwhile, the alleged victim has made no comment, and a grassroots support movement for the defendants continues to grow.

## *Feature Organizational Strategies*

You can organize a feature many different ways. Longer features almost always begin with a lead (one or two paragraphs) that entertains and hooks readers. These leads don't reveal the full subject, or topic, of the story. That's the job of the "nut graf" or "nut paragraph." A nut graf isn't a feature story's first paragraph, but it does come early in the story—usually right after the lead paragraph or paragraphs. The nut graf tells what the story is really about. It links the lead to the big idea, the main point, of the story. A nut graf is sometimes called a "swing graf" because it swings the lead into the true focus of the story. A nut graf comes early in a feature because you don't want readers wondering too long about the exact subject of the story.

In short features, consisting of only a few paragraphs, the nut graf is sometimes part of the lead paragraph or one of the first few sentences. A first sentence or two hooks readers; the next sentence can then declare more precisely what the story is about.

After the lead and nut graf comes the body of the feature story, which delivers most of the *who, what, where, when, why* and *how* information in an entertaining or dramatic manner.

The conclusion of a feature often is just one or two paragraphs. Unlike the conclusion of an inverted pyramid news story, which can be cut for space, a feature conclusion is indispensable. It may be the most dramatic moment of a story. Feature conclusions, often quotations, generally sum up the story, dramatically noting what it all means or what has been learned. Sections on feature leads and conclusions appear below.

Seven good organizational strategies for features follow. Each uses the "show; don't tell" strategy.

1. **The Gold Coin Theory**. Developed by two great journalists, Roy Peter Clark and Donald Fry, the gold coin theory asks you to imagine that readers fear that your feature story is a dusty, uninviting path that they must walk. Tempt your readers by frequently dropping gold coins onto the path. Gold coins are bits of entertainment; they are fascinating anecdotes, great quotations, incredible facts, amazing visuals, something that makes readers laugh or cry—anything that rewards the readers for reading. Ideally, your readers will think, "I just got rewarded for reading this far. I think I'll keep reading to find more rewards." Sprinkle gold coins throughout your feature. Gold coins *must* exist in the feature's lead and its conclusion.
2. **The *Wall Street Journal* Style 1**. Many *WSJ* features begin with a tightly focused anecdote. Then comes the nut graf. More anecdotes then lead to more information—or more information is followed by illustrative anecdotes. The feature often closes with a quotation or an anecdote that memorably sums up the story's main point.
3. **The *Wall Street Journal* Style 2**. Many *WSJ* features begin with snappy, one-sentence leads. That one sentence gets its own paragraph. The sentence doesn't explain the story. Instead, it's mysterious. Readers keep reading to solve the mystery of the lead.

   Immediately after that snappy sentence comes the nut graf, which announces the focus of the story. From this point on, the organization matches the organization that follows the nut graf in *WSJ* Style 1.

   The difference between these two *Wall Street Journal* styles is the lead's length and snappiness. *WSJ* features offer strong examples of effective organization.
4. **The Magazine Personality Profile**. Magazine personality profiles often begin with an anecdote that tantalizes readers. Frequently, the anecdote describes a crisis that becomes a turning point in the person's life. The goal of the anecdote is to make

readers ask, "How in the world did this person get into this situation? What's the story behind this? And what will happen next? How will they solve this?"

Next comes the nut graf, which gives quick background on the main character and their situation. After reading the nut graf, readers know who the story is about and why this person's story matters.

Next comes the relevant history of the central character. The history helps explain the opening anecdote. It often includes comments from the central character. This history also can include comments from other knowledgeable people. The history moves through time until it reaches the moment described in the opening anecdote. At this point, the central character often comments on the events described in the opening. The history then describes the resolution of the opening anecdote, often with comments from the central character and others.

The conclusion returns to the present and shows readers what the central character is doing now. If the opening anecdote hasn't been completely resolved, the final resolution is presented here. The feature often ends with a dramatic, moving, funny or revealing quotation from the central character. The quotation provides a sense of closure.

5. **The Epic Poetry Strategy**. Magazine personality profiles are based on the epic poetry organizational scheme (remember *The Iliad* and *The Odyssey*?). Epic poetry begins *in medias res*—"in the middle of things," usually at an exciting moment. This leads readers to ask two questions: "How did we get here?" and "What will happen next?"

   The epic then flashes back to the beginning of the hero's story, takes us to the middle (which we already know) and reveals how the hero fared, then moves to a dramatic conclusion.

6. **The Bookend Strategy**. The story begins with a strong, appropriate, compelling image that captures the reader's interest. The story later closes with the same image—but with a twist. The image operates as a set of bookends—those matched props that keep a row of books from falling to the right or left—with one bookend at the beginning of the story and the other bookend, with a twist, at the end.

7. **The Theme Strategy**. The theme strategy resembles the bookend scheme, but instead of just appearing twice, the image is woven throughout the story, including the introduction and conclusion. For example, if you were to compare someone to a fairy tale princess (which probably would be a little trite), a unifying element throughout the story could be references to fairy tale images: handsome princes, dragons, fairy godmothers, witches and so on. Such a technique also is called an extended metaphor. The theme strategy works best with short features; it can get annoying in long stories.

   The theme strategy also resembles the gold coin theory, except that in the theme strategy the gold coins all are thematically related to one another.

## *Feature Lead Strategies*

Feature leads should hook readers, gaining their attention and interest. Several traditional hooking strategies exist:

- A snappy, one-sentence teaser that creates a mystery

  *Example*: "Night after night, the danger came." (What danger? Who was affected? Why did it happen at night? Why did it happen night after night? How dangerous was it?) Readers will continue, wanting to solve these mysteries.

- A short, fascinating anecdote that illustrates the story's main point

  *Example*: "When Mary Smith felt the jagged granite atop Mount McPherson, she knew that she had conquered more than a killer mountain. She had conquered her blindness."

- A fascinating quotation from someone involved in the story

  *Example*: " 'If Mark Edwards walked through that door again,' Tisha Bertram says, 'I don't know whether I'd kiss him or kick him.' "

- An impressive fact (be sure it's truly impressive and interesting)

  *Example*: "Lynn Jones can eat 17 and a half extra-large pepperoni pizzas in 20 minutes."

- A striking image

  *Example*: "The machine squats in a dark corner, wheezing and lurching like a has-been sumo wrestler seeking glory one last time."

- A thought-provoking question that can't be answered with a simple yes or no

  *Example*: "If you couldn't be yourself, who would you want to be?"

Again, feature leads are not as direct as the leads of straight news stories. Because a feature lead doesn't have to directly announce the story's subject, it can develop a hook that compels readers to keep reading.

In feature leads, avoid clichés such as "Little did she know. . . ."

Don't allow the headline of a feature story (or any story) to function as your first sentence. Your feature story should be complete without the headline.

## *Feature Conclusion Strategies*

Many features return to a tight focus in the conclusion—to one person's summary quotation, for example. One good conclusion strategy is to return to the image, question or anecdote developed in the lead and put a new, appropriate twist on it. Readers thus see the lead in a new light and, ideally, understand it even more. Sometimes the conclusion supplies the end of an anecdote that began in the lead. Often, a quotation can supply a dramatic, summary conclusion.

Feature conclusions are dramatic. In the words of the gold coin theory, feature conclusions have a gold coin. The story's most important fact or most entertaining moment may appear in the conclusion. (Again, features do not use the inverted pyramid organization, in which the last paragraph is the least important moment in the story.)

Oddly, a "for more information" paragraph following a feature conclusion doesn't undercut the drama of the true conclusion. Be concise in such paragraphs: "For more information, contact the Center City Humane Society at 555-123-4567."

## Hybrid Stories

A hybrid newsletter/magazine story is a compromise between the straight news story and the feature story. To hook readers, the hybrid begins with a feature-like lead. Then, to save space, it moves to a straight news lead, focusing on the important aspects of *who, what, when, where, why* and *how*. After the feature-like lead, hybrids become inverted pyramids (page 100). Hybrids lack the strong, dramatic conclusions that characterize feature stories. The hybrid story can be a good short form for newsletters.

Hybrid stories generally have a traditional news headline (page 99), but they can have a feature-style headline (page 147).

### Hybrid Story Example

**Palmquist students arrested for damaging grass**

Injuries and penalties are nothing new to football, but several Palmquist University students were caught off guard, so to speak, when campus police cited them for unnecessary roughness to grass. Eight students were arrested Saturday on charges of damaging grass in front of Rogers Hall. The misdemeanor charge could result in fines and a jail sentence.

"The students were incredibly irresponsible," said Charles Poole, clerk of the City Municipal Court. "Campus police say that the grass is badly damaged. But I doubt that they'll pay a fine or go to jail if they're guilty. In cases like this, people usually get sentenced to a certain amount of community service."

Palmquist University President Bertrand Bishop said it was important for current students to remember that they are caretakers of the university for future students.
The eight students, all from Central City, are Derek Barrett, 19; Alec Duncan, 20; Rachel Tobin, 18; Jocelyn Snyder, 21; Kendra Jessie, 18; Maggie-Sophia Smith, 20; Gillian Williams, 21; and Emily Davidson, 20.

## STORYTELLING TIPS

1. **Stay on message**: Be guided by a clear understanding of audience and purpose. Know who your audience is and what its interests in this situation are. Be able to define for yourself the story's strategic (goal-oriented) purpose in one clear sentence.
2. **Do rigorous research**: Gather more information than will ultimately appear in the story. The extra information will give you options and help you see what should be included. Trying to write a story, or any document, without enough research is stressful and frustrating.
3. **Study the competition**: Read many newsletters and magazines. Analyze what works and what doesn't.
4. **Deliver details**: Present specifics, not generalities.
5. **Get dramatic**: Show readers; don't merely tell them. Don't say that a person is busy. Instead, show that they're busy. Let readers draw the conclusion that this person is busy.
6. **Select attribution tense**: Quotation attributions in feature stories usually are in present tense. In straight news stories, they usually are in past tense. In a quotation of more than one sentence, place the attribution after the first sentence. For example, "I rarely speak about this," she says. "It's far too embarrassing."
7. **Transfer the techniques**: Feature-story organization can be used for so-called success stories, or case studies, which agencies sometimes use to describe successful communication campaigns. Success stories can appear in brochures, as inserts in folders and on websites.

# 2s Annual Reports

## Purpose, Audience and Media

In the United States, the federal Securities and Exchange Commission and most stock markets, such as the New York Stock Exchange, require companies that sell stock to issue an annual financial report to their stockholders. Annual reports must feature recent financial information, a year-to-year comparison of financial figures, a description of the organization's upper-level management and a discussion of the company's goals.

Earlier, in The Law and Strategic Writing (pages 55–58), we discussed disclosure law. Publicly owned companies (that is, companies that sell stock) rely heavily on annual reports to meet their legal obligation to disclose financial information. Annual reports, therefore, are serious, fact-laden documents. They may have a glossy, glitzy appearance—many do—but they stick to the facts. The concept of puffery—acceptable exaggeration (page 57)—doesn't apply to annual reports.

Some nonprofit organizations issue annual reports to inform current donors and attract potential donors. However, the law does not require such reports. (Nonprofit organizations do report to the U.S. Internal Revenue Service through Tax Form 990.)

The legally specified audience for an annual report is a company's stockholders. However, the larger audience includes potential investors, investment analysts, financial journalists, employees, potential employees and government regulators. Thus, annual reports sometimes appear to have a split personality: a formal no-frills side and a flashier, friendlier side.

The traditional medium for an annual report remains paper and downloadable PDFs, which often look like glossy magazines, but many companies also offer annual reports as multimedia websites.

**KEY TO SUCCESS**

Annual reports should have a clear theme. They should not hide bad news. They should be specific and present information in a variety of ways, including charts, enlarged quotations and photographs with captions.

## Format/Design

The format of an annual report defies concise description. Again, many annual reports have the format and appearance of a glossy magazine: an attractive cover, a table of contents, sections with titles and dozens of pages with type, photographs and charts.

The mandatory sections on finance and operations often are set in smaller type than are the less technical sections. In paper annual reports, these technical sections often appear on different quality paper from the rest of the report. These differences set the sections apart, helping them to seem more serious than the rest of the report.

In long sections, internal headlines can clarify content and increase readability.

Some companies that sell stock choose to meet stock-exchange reporting requirements simply by posting their annual U.S. Securities and Exchange Form 10-K. Those documents are long, serious business reports that lack attractive graphics. Companies sometimes insert their Form 10-K into larger, more reader-friendly annual reports.

## Content and Organization

Annual reports often have five sections, each of which is discussed below. In general, an annual report should discuss your organization's strategies and performance. The annual report should describe, in terms of fulfilling goals, where your organization is, where it has been and where it plans to be. Be specific. Cite specific goals and precise measurements of performance. Use numbers to show how close to your short-term and long-term goals you are.

Discuss any bad news openly, and show specifically how you're correcting the problem.

If possible, unite all this information with a theme, either implicit or explicit. A theme can help organize the report's information and keep it directed toward a specific message. In other words, ensure that your annual report is on-strategy: Understand what clear message the report should send, and direct all your writing toward the fulfillment of that message and theme. But don't forget that split personality: Even as you try to develop an engaging theme, remember that this is a legal document. You must scrupulously stick to the facts.

The cover title of an annual report traditionally is the name of your organization plus the words "annual report" and the year. Some annual reports also print the explicit theme on the cover.

Work with your company's legal and financial teams to ensure that the annual report complies with federal, state and stock-market laws and regulations.

The following passages describe the traditional sections of an annual report.

### *Opening Charts and Graphs: Basic Financial Information*

These financial charts—such as tables, bar graphs and pie charts—should be clearly labeled and reader-friendly. The charts often have no captions, just clear titles. This short section is typically just one page and often is on the inside front cover. Usually, your role will be to edit the few words that accompany these charts. Financial personnel and the annual report's designer, or art director, prepare the charts.

These opening charts sometimes are combined with the next section: the message from the CEO. Some annual reports place these immediately after the CEO's message. Each of these approaches, however, places these charts near the beginning of the report.

### *Message From the CEO*

This message—sometimes called a letter—focuses on the achievements of the past year and thanks employees, stockholders and other groups that have helped the company work toward its goals. The leader of the company, generally the chief executive officer (CEO), writes this section (though, often, a member of their public relations staff actually writes the section with guidance from the CEO). The CEO's message should clearly reflect their personality. If the annual report has a theme, skillfully integrate that theme into this section.

Within the first few sentences, the CEO should say how the company performed during the past year. Were profits up or down? Why? If profits were down or were disappointing, the CEO should acknowledge that, explain why and discuss what's being done to improve the situation. Investors appreciate—and expect—candor.

The CEO often signs this section—just as if it were a long, informative letter.

### *Longer Section on the Company*

This section resembles a long feature story in a magazine: lively, engaging, well-organized, informative and specific. It may be several related feature stories. This section is one of your best chances to use the annual report to present your company as an attractive investment opportunity. Use smaller, inset articles (called sidebars) and charts and graphs to highlight key information. Coca-Cola, for example, often includes a section on each brand within the company. Photographs with captions also can enliven this section. Not all annual reports include this long section, though most do; sometimes information that would go here can be included in the other sections. And as you've read many times in this section, stick to the facts. Don't exaggerate in this legal document.

### *Management's Analysis of Financial Data*

This section presents financial charts accompanied by long, technical explanations. Financial personnel from your company usually will prepare this section, and it will be verified by an outside accounting agency. You probably won't write this section, but you might help edit those long explanations. Be sure to confirm your edits with your company's legal and financial teams. Some of the financial jargon may seem boring and needlessly complex, but that language often protects your company by complying with laws and other financial reporting guidelines. Casual investors may flip through this section with a yawn, but investment analysts will study it carefully.

## *Who's Who in the Company*

Many annual reports close with information about the board of directors and other high-ranking company officials. This section often simply lists names and titles under individual photographs. This "lite" approach can disappoint readers who want to evaluate the men and women who will implement company strategies. Instead of presenting such limited information, consider including more details in this section. State how long the officials have been with the company and where they were before. List any college degrees they've earned. List the individuals' specific duties, especially as those duties relate to the fulfillment of company goals. Consider quoting the officials on their personal priorities for the organization. Include a group photograph or a photo of each individual—or both.

### ANNUAL REPORT TIPS

1. **Get started early**: Evaluate the previous annual report. What worked well? What didn't? Experts say that you should begin planning seven to eight months before the new report's mailing date. Federal regulations require distribution of the annual report to stockholders at least 40 days before the corporate annual meeting.
2. **Promote—with caution**: Use the annual report to promote your organization as a good investment opportunity. Besides informing stockholders and meeting legal requirements, use your annual report to attract investors and boost stock prices. But remember: Stick to verifiable facts.
3. **Select a design that enhances your words**: Work with the best designer available, one who will use photos, charts and other illustrations to help tell the story. Be sure that the designer has read the current draft of the annual report and knows the overall theme. Some designers forget that their mission is to make the report's message clear and accessible; they become more interested in fascinating design that looks great but doesn't tell the story. The design of your annual report *should* look great—but first, the design must be functional: The design must work with, not against, the words that tell the story.

# 2T Speeches

## Purpose, Audience and Media

In strategic writing, a speech is a scripted monologue designed to be performed in front of an audience. A speech contains a main point—a strategic message—and it elaborates on that main point.

The audience of a speech often is a group that has something that the speaker needs. A presidential candidate addressing a campaign audience needs votes and money. A corporate leader addressing stockholders needs their support. The CEO of a nonprofit organization addressing potential volunteers needs volunteers. Members of the audience usually are united by a common interest.

Speeches can challenge strategic writers because they add a new aspect to the *media* part of our "purpose, audience, media" analysis. Speeches have a speaker—and that speaker often is not the speechwriter. In other words, the speeches you write usually will be delivered by someone else. Therefore, besides studying the purpose of the speech and the audience for the speech, you also must study the speaker. Your script must sound like the speaker at their best—not like you at your best. You must consider the speaker's communication abilities, not your own. In a speech, a complex, unique human being is part of the medium.

Besides the speaker, the medium of a speech can be paper or a display screen, depending on the speaker's preference and the location's technological capacities. Presidential candidates, for example, often use teleprompters. A teleprompter is a see-through screen that allows the speaker to see the speech's words while appearing to maintain eye contact with a camera or an audience. Speeches also can be scrollable documents on tablets and laptops. Because paper is not subject to technological failures, it remains a popular medium for speech scripts. Some speakers prefer to speak from outlines or from notecards.

**KEY TO SUCCESS**

Effective speeches are short, well-organized and focused on the audience's self-interest.

## Format/Design

Format guidelines for speech scripts focus on making the text easy for the speaker to read while maintaining frequent eye contact with members of the audience.

Double-space (or more) between the lines. Use large type and wide margins.

In a paper script, type only on the upper two-thirds of each page so that the speaker's chin doesn't dip too low as they read.

Number the pages of a paper script in the upper-right corner. (That placement will help the speaker avoid accidentally reading the page number aloud.) Put "-more-" at the bottom of each appropriate page. Put "-end-" below the last line of the script. To assist the speaker, avoid page breaks in the middle of a sentence, even if it means moving the entire sentence to the next page.

Don't staple the pages of a paper script—the pages must turn easily and quietly. For a paper script, learn what kind of binding the speaker prefers. Some will want the pages clamped together; they will remove the clamp only when they reach the lectern. Others prefer that the script be three-hole punched and clipped into a narrow three-ring binder. Different lectern sizes may influence the binding of the script.

If you include visual aids, such as PowerPoint slides, note in the script where each new slide occurs so that the speaker can pause and, perhaps, gesture at the screen. Visual aids can be indicated in the script by highlighting relevant passages with a different color; or by inserting bracketed, capital-letter notes; or by placing indicators in the margins of a paper script. Again, learn what the speaker wants.

Include "stage directions" in the script in brackets and, usually, in capital letters. For example, suggest a dramatic pause at a particular point by writing "[PAUSE HERE]" in the script. You also can suggest gestures.

## Content and Organization

Begin by analyzing five things:

1. **The purpose**. After discussing the speech's purpose with the speaker, write, in one sentence, the main point of the speech: the strategic message. Then write a brief description of what the speech should include. Create a working title. Have the speaker review this summary and provide input.
2. **The audience**. Who are its members? What is their common background? What is their strong self-interest in the situation that has prompted the speech? What topics, sure to draw a response, can the speaker discuss? What do audience members expect from the speech? They will be attentive if the speech focuses on their self-interests. Most audience members hope to be informed and, if appropriate, entertained.

3. **The speaker**. Be sure to write *the speaker's* speech—not yours. Again, the speech should sound like the speaker at their best—not you at yours. Spend as much time as possible with the speaker to learn any phrases, gestures and speaking styles.
4. **The time frame**. Has a time length been specified? If not, consider the subject, purpose, audience and speaker in determining how long the speech should be. Whenever possible, limit the speech to 20 minutes or less.
5. **The setting**. Where will the speech be given? Will the location be inside or outside? What is the size of the room? Will there be a lectern and a microphone? Will the microphone be stationary or clip-on? Will audiovisual systems be available if needed? Will a water bottle or glass of water be available?

## *Organize the Speech Logically and Gracefully*

The content of a speech can be organized in many ways. Almost all speeches have an introduction, a body and a conclusion—much like a feature story for a newsletter or a magazine (see pages 145–153).

Most business speeches have a *what* and a *why*: The *what* announces the main point, and the *why* explains or justifies that main point. Some speeches also have a *how* that delivers a call to action; for example, how should audience members ideally respond to a situation that requires action? The *what* and the *why* are the foundations for the following two traditional ways to organize the contents of a speech.

### *Emphasizing What*

Use this organizational strategy when the *what* of the speech is more important than the *why*—that is, when the announcement is more important than the explanation or the justification of the news. Use this strategy when you want audience members to be able to repeat the main point to themselves and others.

1. **Introduction**. Build up to the main point (the *what*) and announce it. If the *what* is something positive that you want to emphasize, make it the last words of the introduction and pause after it.

   You can build up to the main point by thanking audience members for attending and very briefly providing the reasons for the *what*. You could complete this buildup in two or three sentences. If that seems too abrupt, provide a few general comments that provide a transition from the thanks to the reasons.
2. **Body**. Expand on the *what*. One of the best ways to flesh out the main point is to explain what it means to the audience.

   Another way to develop the *what* is to discuss the *who, what, when, where, why* and *how* of the main point.

However you choose to elaborate on the main point, if you have more than one elaboration to make, the subject-restriction-information technique can work well for each new paragraph. This technique shows the relationship of each new elaboration to the main point. For an explanation of subject-restriction-information, see page 331.

3. **Conclusion**. Reiterate the *what* once again and put a dramatic or memorable spin on it.

One effective variation of the *what* organizational strategy is to create a "theme" speech, in which a strong theme is announced in the introduction, along with the *what*. The theme is clearly intertwined throughout the elaboration of the *what* in the body and then dramatically reasserted in the conclusion.

For example, a speech in which an executive announced the retirement of a valued colleague might review the colleague's career and compare it to a year. The theme could include references to spring, summer and autumn. The theme also could include references to planting and harvesting, storms and holidays. It could close by borrowing from an old Frank Sinatra song: "It was a very good year."

## *Emphasizing Why*

Use this organizational strategy when the *why* is just as important as the *what*. Use the *why* organization when you want the audience to understand why something has happened or is about to happen. Use it when you want members of your audience to be able to explain and justify the main point to themselves and others.

Also use the *why* strategy when your main point involves bad news. Just as in bad-news messages (pages 306–310), the *why* organization allows you to give the explanation for the bad news before you actually announce it. Ideally, if audience members understand the *why*, they will better accept the *what*.

The *why* organizational strategy consists of five parts:

1. **Introduction**. The introduction often contains only a greeting. Do not mention the main point of the speech—the *what*—in this brief section. The introduction can be something as simple as "Good afternoon, and thank you for that warm reception."
2. **Explanation**. The body of the speech begins here. Discuss the *why* in this section. Cover the relevant points that explain or justify the main point, even though that main point has not yet been spoken. Basically, establish a cause-effect relationship. This explanation section describes the causes, discussing all the reasons that justify the main point to come. When the speaker then delivers the main point, the audience is prepared to accept it. The listeners already have heard and considered the logic that supports the main point.
3. **Main point**. The body of the speech continues, and the speaker delivers the main point, the *what*. This section usually is brief. Sometimes it consists of only one sentence.

Note again the sequence of parts: The speaker logically builds up to the main point. The main point is not mentioned until the third section of the speech.

4. **Remarks**. As the body of the speech continues, the speaker develops the main point. For example, what are its consequences? What does it mean to audience members?
5. **Conclusion**. The speaker can repeat the main point here, though such repetition often is unnecessary. The conclusion usually is concise. The speaker often appeals to audience members' emotions. Recall how many presidential speeches end with "God bless the United States of America," a powerful appeal to the emotions.

Each part of the five-part *why* speech should lead logically and gracefully to the next.

## *Write for the Ear, Not the Eye*

Audience members can't scan back up the page or hit rewind to clarify meaning. Therefore, consider the following:

- Have a clear beginning, middle and end. However, avoid overworked, boring transitions such as "My next point" and "In conclusion." Use the subject-restriction-information method (page 331) to avoid "My next point." Use the expanded focus of the conclusion (a clear return to a more general tone and to the big picture) to avoid "In conclusion."
- Keep sentences short. However, vary sentence rhythms in accordance with meaning. For example, deliver blunt ideas in short, blunt sentences. Deliver relaxing ideas in longer (though not long), flowing sentences. Avoid using conjunctions such as *and* to unite two sentences into one. Use two sentences.
- Don't use big, pretentious words. Remember the advertising maxim, "Big ideas in small words." It can be done: Recall Hamlet's "To be or not to be—that is the question." He's discussing whether to live or die, and he's using mostly one-syllable words.
- Avoid technical terms unless you're certain that your audience knows and understands them. When you must use unfamiliar terms, define them in clear, simple language.
- In the script, spell out big numbers for the speaker. Consider phonetically spelling tough words or tough names. Learn what the speaker is comfortable with.
- Don't expect the audience to remember more than two or three key ideas.
- Review the tips in Writing for the Ears on pages 29–34.

## SPEECH TIPS

1. **Create a title**: Give the speech a title, even if that title is known only to you and the speaker. A title can help you stay focused on the main point, the strategic message.
2. **Consider an opening hook**: Instead of saying, "Thank you. I'm glad to be here" (or immediately thereafter), wake up the audience with a provocative question, an attention-grabbing statement or a short, entertaining anecdote.
3. **Think twice about jokes**: Many speechwriters recommend that you avoid jokes, especially jokes with punch lines. If the speaker misspeaks the line or if the joke isn't funny—or worse, if it's inappropriate—the speech and the speaker are damaged.
4. **Create images with words**: Be innovative in how you present facts. For example: "We use four tons of paper a day. That's enough to bury this room eight feet deep every working day of the year. Eight feet deep!"
5. **Focus on the audience**: Address the self-interest of audience members. Talk to listeners about themselves.
6. **Consider multimedia**: Use simple visual aids, especially computer-generated, if appropriate and available. Remember that what is both heard and seen is often more memorable than what is just heard or seen. If you use visual aids, make the images simple. Use few words in visuals. If audience members have to interpret images or read long passages, they're not listening to the speaker. Proofread all such materials, and have others do so. Test any related equipment. Know what backup systems are available and be prepared to switch to them. Don't feel compelled to use visuals. In a short, well-written speech, words alone suffice.
7. **Act out the speech**: Present the speech yourself in front of a critical listener before giving it to the speaker. Get feedback. Then, if time permits, consider acting out the speech for the speaker, especially if they are inexperienced. Use the delivery you hope they will use. Ask for their feedback. (It's not always possible to get this much time from busy executives. And it's not always advisable if the executive is a polished, experienced speaker.)
8. **Encourage a rehearsal**: When possible, have the speaker practice the speech in front of you and others they trust. Coach the speaker's delivery and get feedback from others. In some cases, the speaker may be too busy or too embarrassed to rehearse in front of you. If they're embarrassed, consider tactfully pointing out that it's better to get negative feedback from you than from the real audience.
9. **Have a backup**: Be sure to have more than one copy of the speech in the room where the speaker delivers the speech. If the speaker is using a screen for the speech, ensure that a paper copy is present in case the technology fails.
10. **Scout the location for the speaker**: The speaker will be grateful to learn about the lectern, a glass of water, room size, audiovisual equipment and so on. They'll appreciate your extra effort, and you may learn things that help you write a great speech.
11. **See Appendix D**: Appendix D (pages 357–360) includes tips for oral presentations.

## Speech Manuscript

Austin Headquarters Speech–1

Good morning! Thank you for that warm welcome. Before we begin, please look around you. Go ahead . . . look around. [PAUSE AND SMILE.]

We're pretty crowded in here, aren't we? In fact, we're crowded in <u>every</u> room of this great, historic building. We're a growing company. We need our headquarters to grow with us.

Therefore, today I am pleased to announce that our headquarters <u>*will*</u> grow! Last night, we completed the purchase of the old limestone firehouse next door.

-more-

SECTION

# 3

# Strategic Writing in Advertising

## OBJECTIVES

**In Section 3: Strategic Writing in Advertising, you will learn to write these documents:**

- Strategic message planners
- Creative briefs
- Print promotions
- Audio advertisements
- Video advertisements
- Digital advertisements
- Radio and TV promotions
- Public service announcements

# 3A Introduction to Advertising

Advertising consists of persuasive messages sent by identified sponsors to targeted consumers through controlled media.

However, that standard definition just doesn't convey the excitement of advertising. Few things in strategic writing beat the satisfaction of writing a successful ad. Don't you wish you were the creative genius behind Nike's "Just Do It" slogan or Budweiser's Clydesdale and puppy ads? Still, let's examine our definition of advertising a little more. By *identified sponsor*, we mean your client or product. By *persuasive message*, we mean a message designed to influence its recipient. And by *controlled media*, we mean placements (such as websites or mobile messages) in which you can control what your message is, how often you send it, and where and when it appears. Controlled media aren't free, however. As an advertiser, you pay not only to create your persuasive messages but also to place them in the media.

Modern advertising faces severe challenges. We're bombarded by ads in every imaginable medium—phones, laptops, tablets, books, magazines, buses, billboards, even, sometimes, in restrooms. Ads are everywhere. And you expect your message to be noticed in that avalanche of persuasion? A successful ad must cut through that clutter. To win a consumer's attention, you must conduct extensive research on the client, product, competition and target audience. And you must be creative: Everyone can be creative—some just have to try harder. Professors Sandra Moriarty and Bruce Vanden Bergh note that advertising may be the only profession in which some people actually have the word *creative* in their job titles. Perhaps your personal career goal is to become an advertising agency's creative director.

But creativity alone can't guarantee a successful ad. In fact, creativity probably matters less than a related area: the research you conduct. Your research should lead you to the development of one clear message for your ad. (If your ad has several messages or even one unclear message, how can it hope to fight its way through the clutter of competing ads and get results?) We call this one clear message a strategic message. The one clear message is strategic because it focuses on the precise goal of the advertisement. As we've said before, strategic means goal oriented.

Two of the most important documents you'll encounter in this section on advertising are the strategic message planner and the creative brief. Different advertising professionals have different names for these documents, including strategy statements and creative platforms. No matter what we call them, strategic message planners and creative briefs help you organize your research to create an ad's one clear message: the strategic message.

A strategic message usually focuses on a benefit (ideally, a unique benefit) that your product offers to the target audience. David Ogilvy, an advertising genius of the twentieth century, said that the most important sentence in his classic book *Ogilvy on Advertising* was this: "Advertising which promises no benefit to the consumer does not sell, yet the majority of campaigns contain no promise whatsoever."

Benefits are good things that members of your target audience believe your product will do for them. So your toothpaste creates whiter teeth? That's just a product feature. But to your target audience, that feature means sex appeal—a benefit. So your lawn fertilizer guarantees no weeds? Another product feature. But to your target audience, that feature means freedom on the weekends—no more digging up dandelions. Benefits derive from product features.

Not all product features generate benefits. A product feature that appeals to you may have no value whatsoever to your ad's target audience. Only careful research can tell you which features can create benefits—and which benefit is so important that it belongs in your strategic message.

Benefits can build brands. A brand is a consumer's image of a product (a MacBook Air laptop), a product line (Macintosh computers) or a company (Apple Inc.). As advertisers, we certainly have our own definition of our product's brand, but the consumer's opinion is all that really matters. Brands help differentiate competitors. A brand gives your product a position in a consumer's mental map of a market. For example, which do you prefer: Coke or Pepsi? Nike or Reebok? Burger King or McDonald's? Why? What is your image of each of those products? For you, that image is the brand.

Although benefits usually are essential to successful advertising, they occasionally play only a small role—or no apparent role at all—in so-called image advertising. Generally, image advertising—also known as identity advertising and reminder advertising—promotes a brand whose benefits are so familiar that they don't need to be restated in the ad. Instead, the ad simply reminds consumers that the product exists. Consumers then automatically supply the well-known benefits and the positive brand image. Products that flourish with image advertising include market leaders in the soft-drink and athletic shoe industries. Also, announcement advertising—ads that emphasize only product names and prices—fill newspapers every day. In this book, however, we'll focus more on the challenge of creating ads with benefits-related strategic messages. Page 205 List different creative ways to build a story around a strategic message, including, testimonials and comparisons.

Once you know the research-driven strategic message of your ad, you can begin thinking of images, sounds, headlines, slogans—all the aspects of creativity that make advertising so much fun. We think that you'll find strategic message planners and creative briefs so valuable that you'll soon consider them fun as well.

# 3B Social Media in Advertising

## Purpose, Audience and Media

Welcome to one of the fastest-growing and fastest-evolving areas of strategic writing: advertising in social media. The purpose of social media advertising is to build financially beneficial relationships with individual consumers, often by persuading them to follow an organization's social media posts and to click or tap to specific landing pages within an organization's website. Amid these rapid changes, however, the Pew Research Center reports that audience size and platform preferences have remained stable for several years. YouTube remains the most visited social medium, with 73% of U.S. adults reporting usage of that platform, followed by Facebook (69%), Instagram (37%), Pinterest (28%), LinkedIn (27%), Snapchat (24%), Twitter (22%), WhatsApp (20%) and Reddit (11%).

Other portions of this book, primarily Microblogs and Status Updates (pages 70–75) and Digital Advertisements (pages 217–224), will cover best practices for creating social media ads. The purpose of this brief portion will be to describe the social media advertising environment, primarily the growing importance of such advertising; the different advertising specifications of the different platforms; and challenges to success, such as ad-blocking programs.

## The Growth of Social Media Advertising

The website Social Media Examiner recently surveyed almost 5,000-plus marketing professionals in the United States, the United Kingdom, Canada, Australia and other nations and gathered the following advertising data:

- 72% of respondents represented organizations that advertised on Facebook; 59% said they planned to increase their amount of Facebook advertising.
- 38% of respondents represented organizations that advertised on Instagram; 55% said they planned to increase their amount of Instagram advertising.
- 14% of respondents represented organizations that advertised on LinkedIn; 35% said they planned to increase their amount of LinkedIn advertising.
- 12% of respondents represented organizations that advertised on YouTube; 40% said they planned to increase their amount of YouTube advertising.

- 9% of respondents represented organizations that advertised on Twitter; 21% said they planned to increase their amount of Twitter advertising.
- 8% of respondents represented organizations that advertised on Facebook Messenger; 33% said they planned to increase their amount of Facebook Messenger advertising.
- 4% of respondents represented organizations that advertised on Pinterest; 17% said they planned to increase their amount of Pinterest advertising.
- 2% of respondents represented organizations that advertised on Snapchat; 8% said they planned to increase their amount of Snapchat advertising.

Worldwide spending on social media advertising surged past spending for print advertising in 2019, placing it behind only television and paid-search advertising expenditures. Worldwide spending for social media ads is expected to rise to almost $134 billion by 2024.

## Platform Advertising Requirements

Social media advertisers should familiarize themselves with the advertising-guidance sections of each social media platform. Facebook, for example, has Facebook for Business; Instagram has Build Your Business on Instagram; and Twitter has How Twitter Ads Work. Those user-friendly sites can help you design and refine a social media ad—and each can lead you through the specific requirements for each medium. Another handy source is Sprout Social's aptly named "Always-Up-To-Date Guide to Social Media Image Sizes." Did you know, for example, that Facebook profile photos must be at least 180 x 180 pixels but will display on laptops as 170 x 170 and on phones as 128 x 128? The Sprout guide takes you deep into the details for Facebook, Twitter, Instagram, LinkedIn, Pinterest, YouTube and Tumblr.

## Ad-Blocking Programs

A threat to many forms of digital advertising is the growth of ad-blocking programs, which can prevent the appearance of ads in individual users' views of websites and, to a lesser degree, in social media platforms. GlobalWebIndex reports that one-third of internet users have installed ad-blocking programs, and CNET reports that one-third of us have loaded ad-blockers into our phones. Data analytics, such as counting the number of clicks or taps on an ad, can help advertisers gauge the visibility and effectiveness of their ads. Additionally, companies such as Blockthrough offer assistance in designing ads that can counteract ad-blocking software. However, the ad-blocking wars (everyone from *Mother Jones* magazine to the Federal Trade Commission now uses that term) promise to be both ferocious and innovative in years to come.

Let's end this on an optimistic note: Hootsuite reports that 90% of Americans ages 18–29 use social media. The average social media user has nine accounts and uses social media for more than two hours every day for everything from entertainment to shopping to work to keeping up with friends and much more. This is good news for advertising and for strategic writers.

# 3c Strategic Message Planners

## Purpose, Audience and Media

A strategic message planner (SMP) helps create the one, clear strategic message that is the heart of a successful advertisement or advertising campaign. An SMP helps you summarize and study your previous research in order to discover an idea that will motivate the advertisement's target audience to take the desired action. Before strategic writers begin to think of an ad's visuals or jingles or slogans or headlines, they first conduct research and complete a strategic message planner. Why? Because all the creative elements of a successful ad must support one, clear core theme—a strategic message—that unites each element of the ad. Experienced advertising copywriters know that they waste time and money if they try to create an ad before they have conducted extensive research and developed the strategic (goal-oriented) message of the ad. A strategic message planner helps create that core message.

Strategic message planners also are called copy platforms, creative work plans and strategy statements. No matter what you call it, an SMP helps you create your ad's one clear message—the strategic message—by enabling you to organize and study your research.

SMPs generally range in length from four to 10 single-spaced pages. The target audience of your ad or campaign does not see the strategic message planner. Rather, that planning document is created by your creative team and ultimately approved by the client.

Strategic message planners exist either on paper or as online documents that can be easily transferred among members of a creative team.

**KEY TO SUCCESS**

You must complete a strategic message planner *before* you begin to consider the more creative aspects of an ad. (In some cases, you may substitute a creative brief (pages 192–195) for an SMP.) If you begin to create the ad before you develop the ad's strategic message, your creative ideas may bias your idea of what the strategic message truly should be.

## Content and Organization

Brace yourself for a weird fact: Writing SMPs can be fun. Writing a successful SMP is like solving a mystery: *The Mystery of the Unknown Strategic Message*. As an SMP writer, you're a detective. You gather evidence and follow clues. For example, if you learn that your *product* unavoidably is purple, you now have a clue to chase in a different area of your SMP: How does your *target audience* feel about the color purple? Or, conversely, if you learn that members of your *target audience* are passionate recyclers, you need to learn how much of your *product* is recycled and recyclable—and how that compares with *competing products* that your target audience might purchase instead of yours.

If your SMP is written by a team, you'll need to communicate frequently with one another. For example, the writer who is gathering *product* information will need to ask the writer gathering *target-audience* information how those individuals feel about the color purple.

Gathering all this information and chasing all these clues ideally will lead you, the detective, to the mystery's solution: the one, clear strategic message that will fulfill your ad or campaign goal.

So where do we start?

Begin the strategic message planner with a headline that identifies the product: *Strategic Message Planner: Product Name*.

A strategic message planner should be concise, detailed and specific. Before completing an SMP, a strategic writer should conduct extensive research in several areas: client, product, target audience, marketplace and competition. Completing a strategic message planner involves summarizing your research or drawing conclusions in 12 areas:

1. Advertising (or Campaign) Goal
2. Client: Key Facts
3. Product: Key Features
4. Marketplace Trends
5. Target Audience: Demographics, Psychographics and Behavior
6. Product Benefits
7. Direct Competitors and Brand Images
8. Indirect Competitors and Brand Images
9. Product Brand Image (current image, desired image and related challenge)
10. Strategic Message: The Promise
11. Supporting Evidence: The Proof
12. Tone

Each of these subheadings should be a concise, labeled section of your strategic message planner. In each section, consider using bullets or numbered lists, rather than long narratives, to highlight specific, separate facts.

If your SMP is digital, you can link to the sources of information you've gathered. Otherwise, you might consider adding a final References section with numbered sources. At the end of each fact, you could put the number of the source in parentheses.

Especially in the sections about client, product, target audience and marketplace trends, ask yourself, "What is the coolest, most interesting thing I've learned here? What surprised me?" In an SMP, we capture insights, not just information. Those key insights and unexpected discoveries ideally lead you to a powerful, original strategic message—and then to an effective ad or campaign. As you consolidate your research into an SMP, you should take pride in realizing that, probably, you now know more about this exact advertising situation than anyone else in the world.

Let's look at each section of the SMP in more detail.

## *Advertising (or Campaign) Goal*

According to your client, what is the goal of the soon-to-be-created ad or campaign? Of course, the ultimate goal usually will be "to get people to buy the product" or "to make a lot of money."

For this section, avoid an easy answer like that. It won't help you. Instead, focus on *how* the ad will help create those sales and additional profits. For example, the Advertising Goal for a diet product might be "To persuade busy moms that delicious *flavours!* diet meals help give them control over their busy schedules and diets." (See Tip 2 on page 187.)

If your ad achieves that goal, the final goal of creating sales and making more money should be much easier to achieve.

Another way to express a concise ad goal can be to ask, "What problem will this ad try to solve?" Stated differently: "What does success look like?"

The position of a product or client within a particular market can help determine an Advertising Goal. For example, market leaders rarely engage in negative or comparison advertising unless successful rivals are threatening their dominance. Conversely, companies that trail their marketplace competitors are, traditionally, more inclined toward comparison advertising. Sometimes, the discoveries you make in gathering research for your SMP will lead you to revise (with your client's permission) the Advertising Goal.

Your entry for the Advertising Goal section should be very brief—ideally, just one sentence or sentence fragment.

## *Client: Key Facts*

Recent studies show that upward of 80% of consumers say that a company's reputation—its ethos (page 44)—can influence their purchase decisions.

Your client usually is the entity that has hired you to create this ad or campaign. The client is often the producer of the product for which you're creating the ad or campaign.

Given your Advertising Goal, what aspects of your client might help you create a successful advertisement? (And what aspects might be troublesome?) How old is your client? Has it won awards? Is it a socially responsible organization? In particular, what facts might distinguish your client from its competitors? Does your client currently communicate with your target audience? If so, how? What media does it use?

Where can you find such information? Your client, with its website and archives (if any), will be a good place to start, but you should also consult external sources, including news aggregators and databases, such as Google News and LexisNexis, and company-information databases, such as Business Insights, Dun & Bradstreet and Market Share Reporter. A good reference librarian often can recommend top databases for your research.

Consider programming Google Alerts or a similar app to send you an email whenever new stories about your client emerge.

Applying the 13 questions listed in the next section (Product: Key Features) to your client can help you identify key facts that might help shape your ad's strategic message.

Remember that in writing an SMP, you're a detective. At the end of this section, add a concise passage titled **Key Insight**. In this brief conclusion, specify the most interesting, entertaining, useful or unusual fact that you learned.

## *Product: Key Features*

Completing this section involves more than just writing the name and price of your product. Advertising copywriters suggest that, in the research stage that precedes the strategic message planner, you answer the following 13 questions to supply information for this section. You probably won't include *all* your product research in the SMP, but answering these 13 questions may lead to discoveries that can help you create a successful strategic message: Again, keep your eyes open for surprising "I didn't know that!" discoveries. Use your Advertising Goal to help you identify which product features might be useful in the creation of a strategic message.

### *What Is the Product?*

1. To what product category does this product belong? (Product categories can include areas such as soft drinks and athletic shoes.)
2. What are the features of the product (including price and packaging), particularly those features that might distinguish it from other products in the same product category?
3. What specific attributes of the product are discernible through the senses: seeing, hearing, touching, tasting and smelling?

### *What Is the Purpose of the Product?*

4. Why did the organization create the product (besides making money)? What consumer problem(s) does it solve?
5. Have any unintended uses for the product been discovered?

### *What Is the Product Made of?*

6. What are the materials, or ingredients, of the product? Why are those the ingredients? What do they do?
7. Do those materials, or ingredients, have their own ingredients?
8. For services, are there intangible ingredients (such as a particular workout routine at a health club)?

### *Who and What Made and Distributed the Product?*

9. What specific individuals created the product? What are the features of those people, especially features that distinguish them from others in their profession?
10. What vendors supplied materials for the product? What are the features of those vendors, especially features that distinguish them from competitors?
11. What processes did the organization use to create the product? What are the features of those processes, especially features that distinguish them from others?
12. What equipment did the organization use to create the product? What are the attributes of that equipment, especially features that distinguish it from other equipment?
13. Where is the product sold?

In some cases, this Product section may seem to overlap with the previous Client section. A product generally is either a tangible good, such as a video-game console, or a service, such as an airline or a carwash. With services, the distinction between client and product can blur. In such cases, try to separate the history, organization and business dealings of the client from the features of the service that consumers would purchase.

Just as you did with the previous section, close this section with a Key Insight.

## *Marketplace Trends*

Your product exists within a product category. What trends are shaping product development and product buying/selling within that particular marketplace? For example, are cheaper imports dramatically gaining sales? Are new environmental regulations creating the need for product redesign? Has a competitor's product failure cast suspicion on the entire product category? Does overall demand within this marketplace exceed supply?

Identify the major trends, and devote a concise, bulleted entry to each one. Databases such as Mintel, IBISWorld and LexisNexis (and good reference librarians) can steer you to useful information. At the end of this section, add a concise Key Insight that specifies the most interesting, entertaining, useful or unusual fact that you learned.

## *Target Audience: Demographics, Psychographics and Behavior*

Your client or your boss will generally specify the target audience for the upcoming ad or ad campaign. The client will often supply audience research. However, you still must dedicate yourself to understanding the members—the human beings—of that target audience. Sometimes, your additional research may suggest an effective segmenting (a narrowing down) of the original target audience.

Many SMPs begin with a concise summary of the target audience specified by the client—and then use bullet points to add information that provides even deeper insights to that basic group. With the approval of your supervisors and client, you can use your new research to modify that opening paragraph, refining the particular segment of the original target audience that your ad or campaign now will address.

Be sure to avoid this beginner's mistake: *Don't equate the entire target market of the product with the target audience for the specific ad or campaign you plan to create.* In other words, don't try to create one ad that will appeal to all the different groups that might use your product. For example, consumers of milk can include babies, senior citizens and everyone in between. No single ad could successfully appeal to everyone in such a broad target market. Instead, your ad might focus on mothers who buy milk for their teenaged children. You can try to reach the other markets (except, perhaps, for babies) with different, complementary ads.

To define and understand a target audience, you should seek:

- Demographic information
- Psychographic information
- Behavioral information

Demographic information is "nonattitudinal" information—things such as age, gender, race, income and education level. A comprehensive demographic profile also includes the target audience's geographic location(s),

Psychographic information is "attitudinal" information—things such as lifestyles, general attitudes and values, including goals and fears. What things interest your target audience? Some of the target audience's feelings about the product, which also are part of a psychographic profile, will be included in the later section on brand image.

Behavioral information describes your audience's known actions—things such as how often and where they shop (in stores or online). Be sure to understand where members of your target audience are in what marketing professionals call the consumer life cycle: lack of awareness of the product, awareness, pre-purchase research, purchase, repurchase, loyalty or advocacy. Behavioral information also includes purchase history (if any) with the product. For example, advertisers often study a consumer's "RFM" data, meaning *recency* (when did the consumer last purchase the product?), *frequency* (how often does the consumer purchase the product?) and *money* (how much does the consumer spend in each purchasing episode?). Another important area of behavior is media usage: What forms of media are popular with your target audience? Researchers often describe a behavioral analysis as a study of a target audience's habits.

In marketing, the process of using demographic, psychographic and behavioral information to narrow down and deeply understand a particular target audience is called *segmentation*.

Again, your client or your own organization probably will supply research on the target audience. However, if you add your own research (generally a good idea), you should begin by seeking secondary research, which is data gathered by others. Good sources of secondary research include the U.S. Census Bureau, the U.S. Bureau of Labor Statistics, Gallup and the Mintel database. Companies such as Strategic Business Insights, with its VALS analyses (values and lifestyles), can, for a fee, provide target audience research. Another for-profit supplier of marketing-segmentation information is Claritas, with its analyses of consumer segments within every U.S. ZIP code. Again, a reference librarian can suggest additional sources of consumer information.

You also should gather primary research, which is original research that you conduct yourself. Basic primary research methods include in-depth interviews, focus groups, surveys and intense observation, also known as ethnographic research. In addition to a reference librarian, a good textbook on marketing research methods can steer you to reliable secondary sources and teach you how to conduct your own primary research. Just asking questions of people who use your product can lead to important discoveries. In fact, Marketo, which is part of Adobe's marketing services operations, recommends interviewing members of the target audience to create buyer personas—short narratives that, in particular, specify basic demographics, likes and dislikes, preferred media for information and entertainment, and causes of mental pain or stress. Buyer personas can become increasingly detailed as an organization builds its relationships with target audiences.

## *Product Benefits*

A benefit is a product feature that appeals to an ad's target audience. Not all features, therefore, are benefits. In the previous three sections, you studied the features of your client and product and the characteristics of your target audience. You're now equipped to combine your research and determine which product and client features will appeal to members of your target audience. If you list more than one benefit, which benefit would your target audience see as most important? Probably, you will introduce some or all of those benefits in your finished ad.

The one underlying concern of all consumers is "What's in it for me?"—also known as the WIIFM philosophy. In a classic 2010 *AdAge* article, Brian Martin, founder and CEO of the marketing agency Brand Connections, asserted that humans are motivated by 10 basic desires:

1. To feel safe and secure
2. To feel comfortable
3. To be cared for and connected to others
4. To be desired by others
5. To be free to do what they want
6. To grow and become more
7. To serve others and give back
8. To be surprised and excited
9. To believe there is a higher purpose
10. To feel that they matter

Martin's 10 basic desires can be consolidated into three main human needs:

1. The need for control
2. The need for companionship
3. The need for confidence

As you identify the benefits of your product, be certain to consider the consumers' need for control, companionship or confidence. Remember that a key question regarding any product is "What problem(s) does this solve?"

Products that appeal to the need for control can create these benefits:

- Save time
- Save money
- Simplify a task
- Are easy to use
- Eliminate unpleasant tasks or consequences (escape pain)
- Produce pleasure

- Alleviate guilt
- Alleviate fear

Products that appeal to the need for companionship can create these benefits:

- Improve appearance
- Increase sexual attraction
- Increase acceptance and belonging
- Are fashionable
- Improve family relationships

Products that appeal to the need for confidence can create these benefits:

- Lead to praise and accomplishments
- Improve skills and knowledge
- Lead to personal advancement
- Improve personal dependability
- Improve status and protect reputation
- Give pride of ownership
- Give special privileges and recognition

List each product benefit as both a feature and a benefit. For example, you might write this benefit for a diet product: "*flavours!* packaging is made from 100% recycled materials and is recyclable and biodegradable: You're protecting the environment."

*Do not* introduce new facts about your client or product in this section. Any features that you identify as benefits should appear earlier in Client: Key Facts or Product: Key Features.

## *Direct Competitors and Brand Images*

Earlier, when you conducted product research, you placed your product in a product category. To identify direct competitors, name the leading products in that category. For example, if your product (or client, in this case) is Burger King, you certainly would mention McDonald's in the category of burger-oriented fast-food restaurants. For each direct competitor you name, include:

- A concise description of that product, including price
- Your target audience's history with that product
- Your target audience's brand image of that product

- Most important, specific comparisons with advantageous or disadvantageous aspects of *your* product. For example, if your product has recycled packaging and you know that your target audience is environmentally conscious, you need to describe the packaging of each competitor.

Brand image is your target audience's impression of and relationship with a product, a product line or an organization (see page 168). One goal of the strategic message planner is to help you design a brand image that will distinguish your product from its direct competitors. Strategic communicators call this *positioning*: Ideally, the brand image of your product has a beneficial position within the minds of your target audience, especially in comparison with the different positions of competing products.

Your client will often provide information on direct competitors. Brand-image information within particular target audiences can be hard to acquire: It often is established by organizations through private, proprietary research. However, good sources for such information include the Market Share Reporter database and the North American Industry Classification System, which is part of the U.S. Census Bureau database. Don't hesitate to ask members of the target audience whom they see as direct competitors. (And, as you know by now, a good reference librarian can suggest additional sources.)

## *Indirect Competitors and Brand Images*

Completing an entry for this section can be more challenging than describing direct competitors. Indirect competitors are in a product category different from that of your product. However, they are things that could keep your target audience from buying your product. For example, if your product/client is Burger King, indirect competitors might include a leading seafood fast-food chain and even frozen burgers from a supermarket. Specifying brand images for indirect competitors occasionally can be vague and challenging, but you still should make the effort. For example, if you list dieting as an indirect competitor of Burger King, how would your target audience describe its brand image of dieting? And, again, don't hesitate to ask members of the target audience whom they see as indirect competitors.

For each indirect competitor you name, include:

- A concise description of that competitor, including price, if any
- Your target audience's history with that competitor
- Your target audience's brand image of that product
- Most important, specific comparisons with advantageous or disadvantageous aspects of *your* product. For example, if your product has recycled packaging and you know that your target audience is environmentally conscious, you need to describe the packaging of each competitor.

## *Product Brand Image*

This section specifies three related but distinct ideas: current brand image, desired brand image and brand image challenges.

### *Current Brand Image*

Brand image is your target audience's impression of and relationship with whatever it is you're advertising. For example, what do you think of when you hear the name Coca-Cola? Nike? Facebook? Your quick, personal descriptions of those brands are their brand images. Brands can exist at the product level (the MacBook Air), the product-line level (Macintosh computers) and the company level (Apple Inc.).

In describing current brand image, be sure to avoid wishful thinking. Don't describe what you wish your target audience believed. Instead, concisely describe brand image (impression and relationship) as your target audience would. Your research should allow you to do so.

The current brand image often presents a problem. Organizations generally advertise because they believe consumers don't know or don't understand their product. Many times, your advertisement will try to clarify or improve the current brand image and solve the problem. If your product is new, it may not have a brand image with your target audience.

Sometimes, however, the current brand image is exactly what an organization desires. In that case, your advertisement will try to reinforce the current brand image rather than change it. Reinforcing ads often are called image, identity or reminder ads (see page 168).

Your description of current brand image should be brief—just a few sentences at most. Most consumers don't spend time expressing lengthy, highly detailed brand images.

### *Desired Brand Image*

Your client usually will have specific ideas for you regarding desired brand image. Desired brand image is the impression that you wish the target audience had of your product. If your ad is successful, desired brand image would be your target audience's new (or reinforced), enduring impression of the product. Desired brand image isn't necessarily a slogan. Instead, desired brand image is a description. It's how you wish a member of your target audience would, in their own informal words, briefly describe their impression of and relationship with your product. Like the previous section, your entry for this section should be brief—ideally just a few sentences at most.

### *Brand Image Challenge*

What will be the No. 1 problem in moving the target audience from the current brand image to the desired brand image? Perhaps your target doesn't know about the product.

Or, worse, maybe it has tried the product and hasn't liked it. Or perhaps negative publicity has damaged the target audience's relationship with the product. This brief section clearly and concisely describes the top obstacle to your target audience's journey from current brand image to desired brand image. Clearly, your ad will somehow address and try to eliminate this obstacle.

For advertisements that strive to reinforce current brand images, there might not be a brand image challenge.

## *Strategic Message: The Promise*

This section is very important. The words you write here will be the one, clear message of the ad—the theme of the ad. In this section, you clearly state the strategic message that will help generate all the creative elements of the ad. In the book *Creating the Advertising Message*, Jim Albright says that one of the best ways to concisely state the strategic message is to finish this sentence: "Target audience, you should buy this product because __________." The words that follow *because* are the ad's strategic message.

Or you might prefer the approach of David Ogilvy, founder of the agency Ogilvy & Mather, who said, "Advertising that promises no benefit to the consumer does not sell." Ogilvy identified strategic messages by completing this sentence: "Target audience, we promise you that this product will __________." Ogilvy's completed sentence becomes the ad's strategic message.

You won't pluck your strategic message out of thin air. Instead, these important words will come from analysis of the previous nine sections of the SMP. For example, the strategic message will, ideally, overcome the brand image challenge and move the target audience from the current brand image to the desired brand image. And, of course, your strategic message will be benefits-oriented.

The strategic message is not necessarily a slogan. In fact, the strategic message might never appear word-for-word in the ad. Rather, it is the guiding principle of the entire ad. Each element of the ad—words, visuals, music and so on—should reinforce the strategic message.

A good strategic message helps build a unique and positive position for your product within the target audience's mind. A good strategic message makes a beneficial claim that no other competing product can make or has made.

To guarantee the strength of your strategic message, you must test it against the previous nine sections by asking these questions:

- Will the strategic message help move the target audience from the current brand image to the desired brand image? Will it help remove the brand image challenge?
- Will the strategic message set the product apart from the brand images of the indirect competitors? Do any of the indirect competitors send this same message? (If so, the strategic message needs revision.)

- Will the strategic message set the product apart from the brand images of the direct competitors? Do any of the direct competitors send this same message? (If so, the strategic message needs revision.)
- Does the strategic message focus on an important benefit or benefits?
- Is the strategic message consistent with the demographic and psychographic characteristics of the target audience?
- Is the strategic message appropriate for any challenges within the current marketplace?
- Is the strategic message consistent with the key features of the product?
- Is the strategic message consistent with the key features of the client?
- Will the strategic message satisfy the advertising goal?

Notice that in answering these questions, you move backward through the strategic message planner. Doing so ensures that the strategic message is on target and is supported by your research and analysis.

Your strategic message should be brief—ideally, one sentence.

## *Supporting Evidence: The Proof*

Imagine that your target audience has seen your strategic message and responds with this statement: "Oh, yeah? Prove it." In this final section of the strategic message planner, you list the facts—often presented as benefits—that support the truth of your strategic message. These bits of proof sometimes are called selling points. No new information is added to the strategic message planner at this point: The evidence should come from the earlier sections of the SMP.

For example, the strategic message for a new diet product might be "New *flavours!* meals put you in control of a tasty, healthy, convenient and socially responsible diet." Your supporting evidence for this message—drawn from earlier sections in your strategic message planner—would show that *flavours!* meals are indeed delicious and can promote healthy weight loss and the packaging doesn't damage the environment. In most cases, your ad would avoid presenting other benefits, such as low prices or availability in local grocery stores. Discussion of those benefits could weaken the focus on your one, clear strategic message: This product puts you in control of an effective, tasty, convenient, eco-friendly diet.

## *Tone*

You know your basic strategic message and the supporting evidence. Now consider the best tone of voice for your ad. If you're promoting a video game that features realistic hand-to-hand combat, the ad's tone might be terse, edgy and explicit. If you're promoting a romance novel, the tone might be breathless, provocative and teasing. Consider specifying three adjectives for the tone of your ad or campaign. If your ad were a person speaking directly to your target audience, what would that person be like? This section should be brief—perhaps a specification of three adjectives and a one-or-two sentence description of the personified ad.

## Format/Design

The format of a strategic message planner (SMP) is simple. It is titled **Strategic Message Planner: Product Name**. It is single-spaced, and it has subheadlines for the 12 sections described previously. For example, the Product Benefits section should be labeled **Product Benefits** in boldface type. Generally, there is a blank line (a double-space) before each new section.

Strategic message planners are concise. They do not include *all* your research regarding the client, product, target audience and competition. Instead, they include only the highlights relevant to the specified advertising goal. SMPs generally are three to five pages long. A good heading for a second page is "Product Name SMP–2."

Some strategic message planners include a title page that specifies the client, the account executive in charge of the ad and the date. If the SMP is printed, there also might be an "Approved" line with a space for the client to initial, indicating their approval of the document.

## STRATEGIC MESSAGE PLANNER TIPS

1. **Investigate before you create**: Always complete a strategic message planner before you begin an ad. (The powerful temptation will be to rush to the design or copywriting aspects.) If you begin to think about images and headlines or music, you reduce your ability to develop an effective strategic message. The strategy should drive the creative elements, not the other way around.
2. **Let the Advertising Goal evolve**: It's OK to begin with an Advertising Goal that focuses simply on increasing sales within a particular target audience. Sometimes that's all that your client specifies. But as you learn more about your client, product, target audience, marketplace and competing brand images, consider refining the Advertising Goal to indicate *how* you plan to increase sales.
3. **Segment the target audience**: Specify just one well-defined target audience for your ad. Don't try to create an ad that will appeal both to college students and to young married professionals. Those two different groups may be target markets for your product, but it's unlikely that one ad can appeal to them both.
4. **Don't echo competing strategies**: Create a unique strategic message. The strategic message should make a claim for your product that no competing product could make or is making.
5. **Coordinate related campaigns**: Realize that when you advertise the same product to a different target audience, the strategic message will change; different target audiences often seek different benefits. Because advertising often is seen by multiple segments, the strategic messages in a coordinated campaign to different target audiences should complement—and certainly not contradict—one another.
6. **Proofread** carefully: The client who reviews your SMP needs to trust your attention to detail.

A creative brief (pages 192–195) is a much shorter version of an SMP. Consider writing a creative brief when there just isn't time to write a full SMP or when you're creating new tactics for a campaign that already has a trusted SMP. That earlier SMP can provide the information for a concise creative brief for the new tactics.

# Strategic Message Planner

Page 1

## Strategic Message Planner: *flavours!*

### Advertising Goal

To persuade busy moms in the United States and Canada that delicious *flavours!* diet meals help give them control over their busy schedules and diets.

### Client: Key Facts

1. Flavours!4U Products Inc. of Morning City, Michigan, makes *flavours!* diet meals.
2. Until 2021, Flavours!4U was an employee-owned company that distributed its products only in four upper-Midwest states: Michigan, Wisconsin, Minnesota and Ohio.
3. In 2021, OneHandle plc, a multinational corporation based in London, England, purchased Flavours!4U. It contractually agreed to allow Flavours!4U to operate as an independent company.
4. In 2021, Flavours!4U began to distribute its products throughout the United States and Canada.
5. The company's primary products are *flaves!* diet drinks and *flavours!* diet meals.
6. Flavours!4U Products now primarily sells its products to grocery stores in the United States and Canada.
7. Total sales in the past fiscal year were $34 million.
8. Flavours!4U was named one of the "200 Best Companies to Work for in the United States" by *U.S. Workforce* magazine every year since 2018.
9. The Association of Minority Group Managers has named Flavours!4U a top company for employee diversity every year since 2013.
10. Emma Becker founded Flavours!4U Products in 2010. Becker is an award-winning gourmet chef who, from 2009 to 2011, lost 40 pounds following the *flavours!* approach. In 2019, she received a Lifetime Achievement Award from the Master Chefs of America Association.
11. Becker is 45 years old. She has been married for 25 years and has two children: a son, 18, and a daughter, 14. On average, she attends seven school-related, children-related events per school-year month, including concerts, plays and sporting events.

#### Key Insight

Founder Emma Becker understands the lives her customers are living. She's been there; she is there. And the CEO is the main chef!

### Product: Key Features

#### What Is the Product?

1. *flavours!* is a diet program that consists of packaged, low-calorie meals.
2. *flavours!* meals contain 350–400 calories.
3. *flavours!* meals feature 34 different entrées for breakfast, lunch and dinner.
4. *flavours!* meals sell for approximately $11 apiece.

## Strategic Message Planner

**Page 2**

**What Is the Purpose of the Product?**

1. The purpose of *flavours!* meals is to provide tasty, healthy, low-calorie meals. "Losing weight shouldn't mean losing your passion for a healthy, delicious meal," said Emma Becker.
2. Quick preparation time for consumers also is a *flavours!* goal.
3. Product testing has shown improved health and significant progress toward weight-loss goals within three months of regular use.

**What Is the Product Made of?**

1. *flavours!* meals feature organic fresh herbs and spices, including basil, cumin, dill, garlic, oregano, paprika, sage and rosemary.
2. *flavours!* meals feature organic fresh vegetables and lean cuts of meat.
3. The meals contain 100% of the recommended daily value for 21 vitamins and minerals needed for a healthy diet: vitamin A, thiamin, riboflavin, niacin, vitamin B5, vitamin B6, vitamin B12, biotin, vitamin C, vitamin D, vitamin E, folic acid, vitamin K, calcium, fluoride, iodine, iron, magnesium, phosphorus, potassium and zinc.
4. *flavours!* packaging is made from 100% recycled materials and is both recyclable and biodegradable.

**Who and What Made and Distributed the Product?**

1. Award-winning chef Emma Becker created the *flavours!* meals. She uses them herself—first to lose weight and now to maintain her chosen weight.
2. Each *flavours!* meal is vacuum-sealed in a pouch and can be heated quickly in a microwave. Vacuum sealing is superior to freezing in preserving taste and flavors.
3. The vacuum-sealing technique keeps herbs and spices fresh and vegetables crisp and tender.
4. *flavours!* meals are primarily sold in grocery stores in the United States and Canada.

**Key Insight**

*flavours!* meals are convenient, tasty, affordable and environmentally responsible.

### Marketplace Trends

1. Sales of pre-packaged diet meals increased 4% from 2020 to 2021, the 19th consecutive year of growth.
2. Organic meals represent 22% of the prepared-meals market, up 2% from the previous year.
3. Body-weight acceptance is increasing. Weight loss remains the top reason for purchasing diet meals (77% of dieters list it as important), but healthy eating (68%) and weight maintenance (63%) have gained in importance every year since 2012.
4. Most diet meals are purchased in grocery stores or health-food stores (84%), but online purchasing has increased 1% per year since 2018.

## Strategic Message Planner

**Page 3**

5. Almost two-thirds (61%) of dieters read ingredients labels. More than half (54%) say that fresh, natural ingredients are important.

**Key Insight**

This is a growing market, and fresh, organic, healthy diet meals are gaining importance.

**Target Audience: Demographics, Psychographics and Behaviors**

The target audience for this ad is women in the United States and Canada ages 24 to 45 who are concerned about their appearance, particularly their weight. They typically are 15 to 50 pounds over their desired, healthy weights and have tried numerous diet programs without lasting success. If they're not on a diet right now, they're about to start a new one. A 2021 Palmquist University study found that regular dieters often feel trapped within boring, unending, ineffective diet plans.

**Demographics**

1. The average age of the target audience is 30.
2. They have a median household income of $85,000. They are first-time homeowners.
3. They are married and have, on average, two young children. They have been unable to return to their pre-pregnancy weights.
4. They are high school graduates, and approximately 70% are college graduates.
5. Members of the target audience live primarily in mid-sized to large cities.
6. Almost all have careers outside the home.

**Psychographics**

1. Members of this target audience describe themselves as "overwhelmed by good things." They love their families and jobs, friendships, social activities and volunteer work, but they feel like they're always 15 minutes behind. They collapse into bed at night and wake up feeling as if they didn't get enough sleep.
2. They describe their lives as "juggling acts" and "always putting out the hottest fire."
3. These women are more concerned with convenience than cost. However, they truly enjoy good food. They refuse to consider a diet of unpleasant drinks or what they call "rabbit food." For them, gourmet food is one of the finer things in life that they aspire to. They enjoy their rare evenings at fancy restaurants.
4. Even though they're pressed for time, they're willing to spend a few extra minutes to prepare one meal for family members and a separate meal for themselves—if the low-cal meal is part of an easy, effective diet plan that doesn't make them feel deprived as they watch family members eat bigger, tasty meals.
5. The target audience doesn't expect a three-week miracle. Its members are willing to invest long-term in a diet that satisfies their desire for good food and for results. They're educated, and they understand that a diet requires an investment of time. Unfortunately, boring, bland diets—or complicated, time-consuming diets—have made it difficult for them to go the distance.

# Strategic Message Planner

**Page 4**

6. Their families are the most important things in their lives, followed by friends and careers.
7. Because of their focus on their children, they are increasingly passionate about environmental issues: They recycle and try to avoid buying plastic or wasteful packaging. They read ingredients labels and strive to avoid artificial ingredients when possible. They are willing to pay a little more for an environmentally responsible product from a company that is a good corporate citizen.

**Behaviors**

1. They attend their children's school events and tend to enroll their children in a variety of programs, ranging from music lessons to swimming lessons.
2. They use social media, primarily Facebook, Twitter, Pinterest and Instagram, for both entertainment and for current events information.
3. They also get radio news while commuting, and they strive to watch at least three national TV news programs per week, primarily CNN, to keep up with world events. They watch local TV news in the morning, primarily for traffic and weather news. A treat is sitting down and paging through a health or home-improvement magazine.
4. They lead busy lives and don't have time or energy for elaborate cooking. They're too busy to coordinate an elaborate, complicated personal diet and exercise program.
5. Most members of the target audience have not tried *flavours!*

**Key Insight**

These women are super-busy. They yearn to feel in control of their lives.

## Product Benefits

1. *flavours!* meals are prepared and microwaveable: You'll save time during family meal preparations.
2. *flavours!* meals have fresh, delicious ingredients: They'll satisfy your desire for the pleasure of gourmet food.
3. *flavours!* meals have 34 different options for breakfast, lunch and dinner: You have choices and will enjoy the variety.
4. *flavours!* meals are proven to be effective: You will lose weight and maintain your desired weight.
5. *flavours!* packaging is made from 100% recycled materials and is recyclable and biodegradable: You're protecting the environment.

## Direct Competitors and Brand Images

Direct competitors are store-bought diet meals.

1. Ms. Svelte Model Diet Meals
   - Ms. Svelte Model Diet Meals is currently the market leader, with 27% of diet-meal sales.

## Strategic Message Planner

**Page 5**

- At 300 calories, this product line offers 10 different meals, each priced at approximately $10. They offer the recommended daily value of 13 vitamins and minerals.
- Two Ms. Svelte meals are organic. None have recycled, 100% biodegradable packaging.
- The CEO is not the main chef.
- Target audience members buy the product only sporadically when it's on sale or they have a coupon. They do not feel loyalty to this competitor and are not regular purchasers.
- The target audience believes that this leading brand could lead to weight loss. However, they wish the meals tasted better and offered more than 10 entrées.

2. Diet Gourmand Entrées
    - Diet Gourmand Entrées ranks second in prepared diet-meal sales, with 19% of the market.
    - At 400-plus calories, these frozen meals offer 12 different entrées at an average cost of $11. They offer the recommended daily value of 16 vitamins and minerals.
    - These meals offer no organic options. Packaging is 100% recycled materials. Biodegradability is not specified.
    - The CEO is not the main chef.
    - As it does with Ms. Svelte Model Diet Meals, the target audience buys this product only sporadically when it's on sale. It does not feel loyalty to this competitor and does not purchase these meals regularly.
    - The target audience believes that these meals are the best-tasting diet meals on the market, but its members find the meals—at 400–500 calories—to be less effective at weight reduction than other diet meals. The 12 entrées average $11.
3. Chart-Your-Day Weight Loss Foods
    - Chart-Your-Day Weight Loss Foods ranks third in prepared diet-meal sales with 8% of the market.
    - This competitor uses a phone and tablet app that charts daily food intake and recommends specific meals that keep users on their diets. In addition to calories, it charts the daily intake of 24 vitamins and minerals. Meals are ordered and delivered in two-week cycles. Monthly fees range from $120 to $400.
    - Forty-two different meals are available. Sixteen meals use organic ingredients. Packaging uses "up to 50%" recycled materials. Biodegradability is not specified.
    - The CEO is not the main chef.
    - The target audience is aware of this program because of its effectiveness but rarely enrolls in the program.
    - The target audience views this system, which involves charts and record-keeping, as effective but too complicated, time-consuming and expensive.

# Strategic Message Planner

**Page 6**

**Indirect Competitors and Brand Images**

1. Eating less and exercising
   - The target audience believes this would be the least costly and most effective diet plan, but it also believes that it lacks the time and willpower to stay with the plan.
   - The target audience believes this option is boring and difficult.
2. WeightBusters Diet Liquids
   - This competitor comes in packages of four 11-fluid-ounce cans. There are four flavors: chocolate, vanilla, banana and strawberry. The price for a four-pack is $16. The packaging is recyclable.
   - The target audience occasionally uses this product to supplement a diet in progress, but its members feel no loyalty to it.
   - The target audience views these canned drinks as effective but unsatisfying. Members believe that the flavors fail to mask a chalky taste.

**Product Brand Image**

- Current brand image: *flavours!* is an unfamiliar product that recently appeared in local grocery stores.
- Desired brand image: *flavours!* respects my busy life and desire for a tasty, low-cal meal.
- Brand image challenge: The target audience lacks incentive to try a new, unknown diet product.

**Strategic Message: The Promise**

New *flavours!* meals put you in control of a tasty, convenient, effective and socially responsible diet.

**Supporting Evidence: The Proof**

1. *flavours!* meals are microwaveable and require no other preparation.
2. *flavours!* meals have fresh, organic ingredients and the recommended daily value for 21 vitamins and minerals.
3. *flavours!* meals have 350–400 calories.
4. *flavours!* meals offer 34 different entrées.
5. *flavours!* packaging is 100% biodegradable.

**Tone**

Understanding, informative and encouraging. Our voice is that of a close friend who's been there, done that and found the solution that she wants to share.

# 3D Creative Briefs

## Purpose, Audience and Media

A creative brief is a shorter version of an SMP, usually just one or two single-spaced pages. Why are there two different versions of this important planning document?

- We often prepare an SMP for an ad campaign that's new to our creative team. SMPs contain much more information than creative briefs. Creative briefs generally consist of short answers to a series of questions. The extensive information within an SMP can lead to deeper creative insights into effective messages.
- Having completed an SMP for a new campaign, your creative team may choose to use shorter, time-saving creative briefs for new tactics in the now-familiar campaign. Basically, the earlier SMP now serves as a research library for the shorter creative briefs.
- Sometimes, a client might be overwhelmed by the depth of an SMP. (Remember that clients generally approve SMPs or creative briefs before the creative team begins creating ads and other tactics.) A creative brief based on an earlier SMP can be a meeting-friendly alternative.
- Occasionally, there isn't time to assemble the large amount of information in an SMP: In those situations, we need to quickly plan and deliver a well-researched message to a specific audience. Creative briefs are ideal for those situations.
- Whether you create an SMP or a creative brief—or both—we hope the bottom line here is clear: You must conduct research before you begin to plan the message and creative elements of a successful ad.

The audience for your creative brief generally is your client and your creative team. Creative briefs can be used to gain your client's permission to proceed with the ad or campaign—and they can keep your creative team focused on the target audience and the strategic message you intend to deliver to that audience. Unlike a finished ad, few people will see or hear your creative brief.

Creative briefs exist either on paper or as online documents that can be transferred easily among members of a creative team.

**KEY TO SUCCESS**

You must complete a strategic message planner or a creative brief before you begin to consider the more creative aspects of an ad. If you begin to create the ad before you develop the ad's strategic message, your creative ideas may bias your idea of what the strategic message truly should be.

## Content and Organization

Like an SMP, a creative brief summarizes research in order to identify a clear, effective strategic message to be delivered to a well-defined target audience.

Just as with SMPs, creative briefs come in many different forms. Most, however, attempt to concisely answer questions similar to these eight:

1. **Why are we communicating?**
   Our answer should be one or two sentences.
2. **Who is our target audience?**
   Our answer should be an information-packed paragraph.
3. **Given our reason for communicating, what's the single most important thing we know about our target audience?**
   Our answer should be one or two sentences.
4. **What does our target audience currently think about our product?**
   Our answer should be one or two sentences. (Products are tangible goods or services.)
5. **What do we want our audience to think about our product?**
   Our answer should be one or two sentences.
6. **What basic promise addressing the target's self-interests can we make to guide our target to that new understanding of our product?**
   Our answer should be one sentence.
7. **What evidence supports our basic promise?**
   Our answer should be three to five bullet points.
8. **What tactics and media should we use to communicate this information? Why?**
   Our answer should concisely describe the tactic or tactics and media we'll use to deliver our basic promise and the supporting evidence to our target audience. Our answer should briefly say why those are the best tactics and media.

Unlike SMPs, creative briefs rarely list information sources via links or a References section. Either that information exists in a previously written SMP, or there simply isn't time to build a References section for the new brief.

## Format/Design

Like SMPs, creative briefs exist on paper and on screens. A creative brief is titled **Creative Brief: Product Name**. It is single-spaced, and it uses boldface type for its questions. Answers to the questions are in regular (not bold) type. Generally, there is a blank line (a double-space) before each new section.

### CREATIVE BRIEF TIPS

1. **Review SMP tips**: Please review the tips for SMPs on page 185.
2. **Proofread—and then proofread again**: Creative briefs are concise and easy to read. Careless typos and misspellings, therefore, are easy to spot. Such errors could damage your reputation with your client and your creative team.

## Creative Brief

### Creative Brief: *flavours!*

1. **Why are we communicating?**
   To introduce *flavours!* as an effective, convenient, tasty diet option for busy moms.
2. **Who is our target audience?**
   The target audience is busy moms with careers who are trying to lose weight or maintain their desired weight. These are women in the United States and Canada ages 24 to 45, median household income of $85,000, who are concerned about their appearance, particularly their weight. They typically are 15 to 50 pounds over their desired, healthy weights and have tried numerous diet programs without lasting success. They describe their lives as "juggling acts." They love their families, jobs, friendships, social activities and volunteer work, but they feel like they're always 15 minutes behind. Because of their focus on their children, they are increasingly passionate about environmental issues: They recycle and try to avoid buying plastic or wasteful packaging.
3. **Given our reason for communicating, what's the single most important thing we know about our target audience?**
   They're super-busy with family and careers and feel as if they're barely in control.
4. **What does our target audience currently think about our product?**
   They may have recently seen it on store shelves, but they don't yet know it because it's new.
5. **What do we want our audience to think about our product?**
   *flavours!* is a tasty problem-solver—not, on the contrary, another complication in their busy lives.
6. **What basic promise addressing the target's self-interests can we make to guide our target to that new understanding of our product?**
   Delicious *flavours!* meals help uncomplicate your life.
7. **What evidence supports our basic promise?**
   1. *flavours!* meals are microwaveable and require no other preparation.
   2. *flavours!* meals have fresh, organic ingredients and the recommended daily value for 21 vitamins and minerals.
   3. *flavours!* meals have 350–400 calories.
   4. *flavours!* meals offer 34 different entrées.
   5. *flavours!* packaging is 100% biodegradable.
8. **What tactics and media should we use to communicate this information? Why?**
   Print ads in health and home-improvement magazines. Digital ads that link to informative landing sites. News releases. Drive-time radio ads. Active social media engagement, particularly with Instagram and Facebook. Point-of-purchase tastings in stores. We need to pursue multiple avenues to gain audience awareness.

# 3E Print Promotions

## Purpose, Audience and Media

Print promotions are paper-based advertisements, coupons, fliers, posters—any paper-based tactic that traditionally includes a headline, a visual and, often, a short section of promotional writing. Many of these promotions also can be digital. This section will focus primarily on print advertisements that include a promotional paragraph or two. The principles that apply to the creation of print ads apply to many other forms of promotional print tactics.

Print advertisements are persuasive messages that appear in newspapers, magazines or other controlled paper media. Their goal is to get a specific audience to take a specified action—for example, to buy a product (a good or a service), make an inquiry, fill out and return a coupon, call a phone number, visit a website or adopt a new brand image of the product. Print ads often are part of a larger advertising campaign that may include radio, television, mobile, direct mail, outdoor and online advertising designed to create a desired brand image.

Print ads fall into three categories: announcement ads, image ads (also known as identity or reminder ads) and product ads. Announcement ads appear in newspapers, mobile messages and e-blasts every day. They are grocery store ads, pop-up ads linked to a previous user query, coupons and Black Friday sales ads. In mobile messages, pop-ups and e-blasts, they link to a merchant's website or landing page. They contain very little copy except for the names and prices of the items for sale. They must be timely and direct to be effective. Image ads remind the consumer of a product or brand. They sustain consumer awareness and depend on consumers' prior knowledge or a larger marketing campaign to deliver the real product message. Industry leaders such as Nike or Coca-Cola use image advertising. Product ads, our focus in this segment, persuade the consumer to take an action or change an attitude or a behavior. Although there is no definitive right or wrong way to create an effective ad, some basic guidelines can help you succeed.

The audience for a print ad should be precisely defined in the strategic message planner or creative brief that you complete before creating your ad (pages 171–195). The audience for a print ad is *not* everyone who might purchase the product. Amusement parks, for example, have different advertising strategies for parents and children. Usually, the target audience is a well-defined segment of the total potential market for the product.

The medium for a print ad is paper, usually in a newspaper or magazine. Many of the principles of successful print ads, however, transfer to digital ads (pages 217–224).

**KEY TO SUCCESS**

Effective print ads capture consumers' attention, interest them in the product and motivate them to take action.

## Content and Organization

Effective print ads deliver a clear strategic message. To fine-tune your message, begin with a strategic message planner or creative brief (pages 171–195). The more time and effort you put into researching your product and developing the SMP or creative brief, the easier your copywriting job will be. The goal of your ad is to bring to life the exact strategic message identified in the SMP or creative brief.

One of the oldest advertising copywriting formulas is AIDA: Attention-Interest-Desire-Action. First you must get your readers' attention; then you must interest them in your product. Finally, you must create a desire within them to take a specified action, such as purchasing the product or adopting a particular brand image.

Let's look at each part of an ad and see how the AIDA formula works. The basic parts of a print ad are:

- Visual
- Headline
- Subheadlines (optional)
- Swing line
- Body copy
- Zinger
- Call to action
- Logo/slogan/tagline
- Mandatories

### *Visuals*

Research shows that two-thirds of print ad viewers see the visual first. That makes it your best opportunity to get noticed. Photographs of people in situations with which the reader can identify have a stronger appeal than illustrations or all text. Your visual must direct the consumer's attention to the ad's message. It must create a mood, establish a theme or tell a story that appeals to the consumer. The visual leads the consumer to the headline and body copy. (Eyetracking research for websites, however, shows that viewers tend to see headlines first, rather than visuals, on websites; see pages 84–93.)

## Headlines

The headline, which is read by approximately one-third of a print ad's viewers, must capture the ad's key message and direct the consumer's attention to the body copy. That's a pretty tall order for a line of type containing approximately eight words. That's right, eight words (or even fewer). You want to write short, direct headlines that demand the consumer's attention.

An effective headline contains a key benefit. It shows the consumer why this product will meet a basic need. Look at the list of basic needs described on page 178. Which of these needs does your product fulfill? What problem does your product solve? Look at the list of product benefits you made for your strategic message planner. Which of these are key benefits for your target audience? The answer to these questions is the message your headline must deliver. The headline answers the question "What's in it for me, the consumer?"

Remember the *flavours!* product from the previous segments on strategic message planners and creative briefs? Let's say that *flavours!* fulfills the need for control by helping super-busy dieters lose weight: The tasty, low-calorie meals can be prepared quickly without adding yet another complex task to the dieter's day. Your benefits-driven headline might be *Life a juggling act? Meals shouldn't be!*

A successful headline is not about how great the product is; it's about how great the product can make you, the consumer. Always appeal to the consumer's self-interest.

An effective headline usually will fulfill one or more of these functions:

- Stop the consumer and get their attention
- Target the primary audience
- Identify the product
- Fulfill a need
- Offer a benefit
- Summarize the selling message
- Speak directly to the consumer (imperative mood works well for this)
- Stimulate interest by using strong verbs
- Lure the consumer into the copy
- Be short and avoid unnecessary words

An effective headline might:

- Ask a question
- Make a claim
- Put a new twist on an old phrase
- Tell a story
- Invite the consumer to do something
- Appeal to emotions and desires
- Refer to a familiar problem

## *Subheadlines*

Subheadlines are optional secondary headlines. They clarify the main headline or divide longer copy blocks into manageable chunks. They can expand on the key benefit, add information or provide the second part of a one-two punch. When they clarify the main headline, they generally are longer than that headline.

## *Body Copy: Swing Lines, Benefits, Zingers and Calls to Action*

The body copy begins with a swing line. This opening sentence (or two) moves—or swings—the reader from the headline into the body copy. It coaxes the reader to continue. It acknowledges the headline but moves attention gracefully to the body copy; it builds a bridge between the headline and the specific benefits to come. Occasionally, the first sentence of the body copy functions more as a subheadline, and the second sentence becomes the true swing line. The body copy is the heart of the ad. It's where the reader is rewarded with specific, beneficial information about the product or service. Only one in seven viewers will make it this far, so your sales pitch better be worth it. In the *flavours!* ad on page 202, the two-sentence swing line is "Grab control of your life! Add *flavours!* to your diet." This passage moves the reader from the headline to the upcoming discussion of benefits.

Body copy has a clear beginning, middle and end. The swing line establishes a beginning, which leads to the benefits in the middle. Here you'll use research from your strategic message planner to explain how your product creates control, companionship or confidence for your target audience. Begin with your strongest selling point. Emphasize the key benefits in a way that makes an emotional connection, not just a logical one. Dramatize the feeling your customers will get from your product. Paint pictures with words or tell real-life stories. Testimonials from satisfied customers work well if there is space.

The zinger is the last or second-to-last sentence of the body copy. It focuses on the strategic message. The zinger is snappy, clever and memorable—it zings. The zinger often is a sentence fragment. It might restate a key benefit or answer a question. It might ask a question or leave the consumer with one parting thought. Your reaction to a witty, on-message zinger written by another strategic writer should be "Oh, I wish I'd written that." A good zinger complements—but does not repeat—the headline or the slogan. In the *flavours!* ad on page 202, the zinger is "A perfect balance for life."

The zinger often is followed by a call to action. This line asks for the sale and tells the consumer specifically what to do next—for example, call, stop by or visit a website. This ending of the body copy makes a final sales pitch and creates a sense of urgency. In the *flavours!* ad on page 202, the call to action is "Add *flavours!* to your Diet."

## *Body Copy Guidelines*

- Be conversational. Write to the average person in your target audience.
- Be positive, not negative.
- Focus on the reader. Use *you*.
- Emphasize the benefits of your product.
- Support facts with evidence. Be specific.
- Overcome objections.
- Use present tense and active voice (pages 11–12).
- Avoid vague words and clichés.
- Avoid *-ing* words.
- Use *italics* for emphasis. Don't underline or use all caps.
- Vary the length and structure of your sentences and paragraphs.
- Create copy with action, rhythm and excitement.
- Use metaphors, alliteration, assonance, rhyme, meter.
- Be convincing. Prove your product is what you claim.
- Ask for the sale!

**16 words that sell**

| | | | |
|---|---|---|---|
| benefit | guarantee | money | results |
| easy | health | new | safe |
| free | how to | now | save |
| fun | love | proven | you/your |

### *Logos, Slogans and Taglines*

A logo generally is a visual identifier, such as Nike's swoosh or Ford's blue oval. A logo may be a graphic, or it may be words in a distinct typeface—or it may be both.

A slogan, or tagline, usually appears near the bottom of the ad, just above or below the logo. A slogan is a phrase that consumers closely identify with your client. In a recent survey, advertising executives selected Nike's "Just Do It!" and Apple's "Think Different" as two of the most successful slogans of the past 60 years. In the *flavours!* ad on page 202, the logo is centered at the bottom and includes the slogan "Healthy Meals 4-U."

### *Mandatories*

Mandatories include items that are required by law, such as copyright symbols, registration marks and fairness statements, such as *Equal opportunity employer* or *Member FDIC*. Agencies have specific usage guidelines concerning size of logos and particular phrasings. In the *flavours!* ad on page 202, a mandatory, in the lower left, is "© *flavours!* Inc."

## Format/Design

Good design attracts and holds the consumer's attention. It amplifies the message and provides direction and order. Since a print ad's visual is seen first by two-thirds of the viewers, its design and placement are crucial to effective advertising. Using the principles of good design discussed in Strategic Design (pages 35–41) will help you create effective, visually appealing ads.

### PRINT AD TIPS

1. **Stay on message**: Be creative, but don't let a fun idea for a great visual or clever headline pull you away from the strategic message you developed in your strategic message planner or creative brief.
2. **Be conversational**: Ads generally are informal, even if that means being ungrammatical. Sentence fragments, if clearly done for style, work well in ads.
3. **Don't repeat yourself**: The headline, zinger and slogan should each develop your strategic message, but each should be unique. Your zinger, for example, will lose impact if your reader already has encountered that idea in your headline.
4. **Edit**: Proofread. Follow the procedures recommended in Research, Planning and the Writing Process (pages 14–18). Even one small error can damage your credibility with your supervisor or your client.
5. **Study others' ads**: Analyze what works and what doesn't work. If you like an ad, take a photo of it or copy it. Your ads need to be original, but it helps to have a collection of ideas that work.

## Print Advertisement

**MEALS SHOULDN'T BE!**

Grab control of your life! Add ***flavours!*** to your diet. Get healthy meals made just for you. All you do is heat and serve. No more wondering what to serve yourself after preparing a family meal.

Meals from ***flavours!*** contain only the finest organic ingredients, specially prepared by our award-winning chefs and seasoned to please the most discriminating palate—yours.

With more than 34 entrées to select from, you'll find dining a delectable experience.

Best of all, ***flavours!*** helps you reach and maintain your weight-loss goals. Each ***flavours!*** meal contains 350-400 calories.

Entrées come with complete serving instructions and include all the ingredients you need for a nourishing, appetizing meal. And cleanup is a breeze: ***flavours!*** packaging is 100 percent biodegradable. Your satisfaction is guaranteed.

• • •

Convenient. Delicious. Simple. Healthy. ***flavours!*** A perfect balance for life.

**Add flavours! to Your Diet!**

www.flavours!4U.com

© *flavours!* Inc.

Credit: Gino Santa Maria/shutterstock.com; Nitr/shutterstock.com/ ria_airborne.shutterstock.com

# 3F Audio Advertisements

## Purpose, Audience and Media

The purpose of an audio advertisement is to motivate the listener to take a desired action. Because of their geographic flexibility and relatively low cost, radio advertisements—also known as *radio spots*—often are used to complement advertising messages delivered in other media. Their real power is to reach an audience in times and places other media can't.

However, this is about more than traditional radio advertising. In addition to receiving "terrestrial radio" over the AM and FM broadcast bands, we also get audio-only programming via satellite, internet and wireless devices. Audio ads may be embedded within a podcast. Despite the evolution of audio technology, its ability to target specific audiences remains unchanged. Each audio outlet has its own distinct demographic and psychographic profile. By using strategic message planners or creative briefs (pages 171–195; 194–197), advertising account managers and their creative teams can craft messages suitable for a variety of well-defined target audiences.

Radio and other audio messages engage one's sense of hearing. They are also linear media, which means that their messages are fleeting: The listener often can't replay or reread them. The strength of audio media is that they are portable and can reach listeners in places traditionally unreachable by other media, such as in a car. The major weakness of these messages is that they engage only one of the five senses (unless, of course, the volume is so loud as to physically hurt the listener). As a result, listeners can be distracted easily. Your challenge as a strategic writer is to develop a message that grabs—and keeps—the listener's attention.

### KEY TO SUCCESS

An effective audio advertisement has a simple message and a clear call to action. It takes advantage of the power of voice, music and sound effects.

## Format/Design

Question: How long should your legs be? Answer: As long as they need to be—within limits. The same holds true for audio spots: Their length should be based on the complexity of the message and the knowledge of the audience. Well-established brands such as Coca-Cola and Nike do not require much explanation. However, it may take more time to introduce a new product. Most audio commercial spots, even those sold within podcasts, are written in standard time lengths: 10, 15, 20 and particularly 30 and 60 seconds. Because multiple scripts may be written for the same product, traffic information should be placed in the upper-left corner of the script. This includes the title of the spot, the sponsor/client, spot length and air dates.

Audio scripts generally take three forms: an announcer continuity (reader), a production script and a one-column hybrid.

### *Announcer Continuity (Reader)*

This is the simplest of the various formats, a one-column script for the person or persons doing the narration. An announcer continuity—also referred to as a *reader*—is often read live—or in real time—on the air. Take this approach in connection with live broadcast events or when you want to associate the product with a strong, popular personality. The script is designed for the convenience of the announcer. It is often placed in a ring binder on a stand in front of the microphone. Write the copy in a large, easy-to-read typeface. Double-space it and do not divide words or key phrases on different lines or pages. Within parentheses, provide the announcer with a pronunciation guide—also known as a *pronouncer*—for unusual or difficult words and names.

### *Production Script*

This is for pre-recorded spots. Pre-recording provides a consistent presentation of the message and takes advantage of a variety of voices, music and sound effects. A production script provides a roadmap for how the audio spot is to be produced. For that reason, it uses a more complex, two-column format. The left column contains production instructions for the use of voice, music and sound effects. The right column contains the script that announcers or actors follow. Write the copy that will be read in a large, easy-to-read typeface. Double-space it. Because of the two-column format, you often will need more than one page. For that reason, do not divide words or key phrases on different lines or on different pages. And, as is the case with announcer continuity, provide a pronouncer for difficult words and names.

### *One-Column Hybrid*

This format is a combination of the announcer continuity and production script formats. It is a popular approach, especially when writing a longer script, such as for a documentary or a dramatic production. It eliminates the need to fiddle around with two-column formatting while still providing both the reader and the producer with sufficient direction.

## Content and Organization

The content and organization of each audio advertisement are driven by a strategic message developed in the strategic message planner or creative brief (pages 171–195). Therefore, the content and organization of audio ads vary widely. However, even with those differences, common elements still exist. In writing your script, be sure to follow the Writing for the Ears guidelines on pages 29–34.

The first step in writing an ad is deciding the creative approach it will take. In a sense, this is your plan of attack. Here are some frequently used audio advertising approaches:

- **Demonstration**: Show the product in action. Of course, that's harder to do using audio than it is with visual media. However, the creative use of words and sound effects can overcome this challenge.
- **Problem/Solution**: This is a clear statement of what the product can do for the consumer.
- **Comparison**: This can be a comparison of before-and-after effects or a direct comparison with a competitor. However, the Federal Trade Commission has ruled that such a comparison must be fair, accurate and documented.
- **Spokesperson**: A quick way to draw attention to a product is to have someone famous endorse it. The FTC has ruled that a spokesperson must actually use the product if they say they do. There's also danger in tying the product to someone who falls out of public favor. (Look up O.J. Simpson or Jared Fogle.)
- **Testimonial**: These spots feature folks similar to the ad's targeted audience who praise the product. Sometimes the "man or woman on the street" will have more credibility than a celebrity does.
- **Image**: The reason a consumer chooses one product over a similar competitor often rests in the image it has projected. (Coca-Cola as "the real thing" versus Pepsi as "the drink of a new generation.") Image advertising can be very effective—but only if the product is already well-known and doesn't have an established image.

Once the writer knows the approach, the next step is to write copy that grabs the audience's attention and keeps it. Your spot needs to stand out from the crowd. There's probably a lot going on around your listeners. You want the audience's undivided attention. One way you can do that is through repetition of key information. If the ad's purpose is to generate business, be sure to repeat the name and location of the store. If the purpose is to get people to contact a business, repeat the contact information. If your ad focuses on a product, mention the product's name at least three times in a 30-second ad. Do the same if the ad focuses on a client, such as a restaurant. For added emphasis, the key points should be among the first and last things the listener hears.

Write in concrete terms that are easily understood by the audience. For example, the audience more easily relates to a "going out of business sale" than it does to a "stock liquidation sale."

It is also important to use the medium effectively. When used to its fullest advantage, audio can create mental images that make even Hollywood special effects gurus jealous. Voice, music and sound effects can transport the listener into a time, place, environment and mood of your choosing. But don't overdo it. Every element in the production should complement the strategic message instead of detracting from it.

Finally, don't forget to sell. The bottom line *is* the bottom line. Your ad must include a call to action. Remember these three strategic questions as you write: What does the audience know? What do you want it to do? And what does the audience need to know to do what you want it to do?

## AUDIO ADVERTISEMENT TIPS

1. **Be anonymous**: There's no personal style in copy writing. The copy should reflect the characteristics of the product, not the person writing about it.
2. **Be honest**: Honest copy that delivers what it promises upholds the product's credibility. It encourages repeat purchases. It also keeps you out of court.
3. **Be conversational**: Write the way the target audience talks. Speak the language of your audience. Use familiar words and phrases.
4. **Write to just one person**: Your copy should speak to each member of that audience, one person at a time. For example, "How are you?" is much better than "How is everyone doing?"
5. **Tell a story**: Consider using storytelling skills to capture the listener's attention. Several of the organizational strategies discussed earlier in Newsletter and Magazine Stories can be effective (pages 148–150). This approach is also effective in conjunction with content marketing tactics (pages 245–246).
6. **Talk to yourself**: Read the copy aloud to yourself. An audio script is written to be read aloud. It may look good on the page. But the only true test of quality is whether it sounds good. Sound it out!
7. **Review**: Refamiliarize yourself with the guidelines and vocabulary for Writing for the Ears (pages 29–34).

## Announcer Continuity (Reader)

Title: Active Moms
Client/Sponsor: *flavours!* Length: 15 seconds
Air Dates: January 1–June 1

Too busy to eat right? You can have a healthy meal and lose weight with flavours delicious diet meals. Made with only the freshest, natural ingredients. Choose from more than 30 low-calorie microwaveable meals. Breakfast, lunch and dinner. Flavours. In your grocer's fine foods section.

###

Credit: ria_airborne.shutterstock.com

## Audio Advertisement Production Script

**Client/Sponsor:** Riverview Community Medical Center
**Length:** 30 seconds
**Air Dates:** June 1–September 1

| | |
|---|---|
| SFX: Sounds of people playing softball. (Establish, then fade) | |
| ANNOUNCER: | It used to be your favorite time of year. But you're banged up and on the bench. It's time for you to step up to the plate! |
| SFX: Sounds of a softball being hit off a metal bat with people cheering. | |
| MUSIC: Upbeat. (Establish, then under) | |
| ANNOUNCER: | It's time to visit Rehab and Sports Therapy at Riverview Community Medical Center. Why go for treatment out-of-town when you can get personal attention from people you know? With Riverview Community Medical Center's team of trained therapists, you'll be back in action before you know it. |
| ANNOUNCER: | Ask your family doctor about Rehab and Sports Therapy at Riverview Community Medical Center. |
| ANNOUNCER: | Get back in the game! |
| MUSIC: (Fade, out at :29) | |

###

## One-Column Hybrid Production Script

**Title:** Game On!
**Client/Sponsor:** Riverview Community Medical Center
**Length:** 30 seconds
**Air Dates:** June 1–September 1

**SFX**:

Sounds of people playing softball. (Establish, then fade)

**ANNOUNCER:**

It used to be your favorite time of year. But you're banged up and on the bench. It's time for you to step up to the plate!

**SFX**:

Sounds of a softball being hit off a metal bat with people cheering. (Establish, then fade)

**MUSIC**:

Upbeat. (Establish, then under)

**ANNOUNCER**:

It's time to visit Rehab and Sports Therapy at Riverview Community Medical Center. Why go for treatment out-of-town when you can get personal attention from people you know? With Riverview Community Medical Center's team of trained therapists, you'll be back in action before you know it.

**SFX**:

Sounds of a softball being hit off a metal bat with people cheering. (Establish, then fade)

**ANNOUNCER:**

Ask your family doctor about Rehab and Sports Therapy at Riverview Community Medical Center. Get back in the game!

**MUSIC:**

(Fade, out at :29)

###

# 3G Video Advertisements

## Purpose, Audience and Media

The purpose of a video advertisement is to motivate the viewer to take a desired action, usually in the form of a marketplace transaction. Because of its sight and sound dynamics, television in particular is considered a prestigious medium and seen as especially effective in building and maintaining a corporate or product image. Television ads also are known as *television commercials* or *television spots*.

However, as noted in the discussion of audio advertising, video technology and media have evolved in recent years. Video advertising is now delivered on four screens: traditional televisions, computers, tablets and phones. These ads are no longer limited to scripted or live programming. Video ads are found in websites and social media, including YouTube videos. And these screens are located almost anywhere from Times Square in New York to, yes, even the privacy of a restroom. Video messages are ubiquitous—which makes them something all strategic communicators must understand and embrace.

For television, three major classifications of video advertising exist. The first is local (or *spot*) advertising for individual stations or cable companies. National/global advertising is distributed to much larger audiences through networked programming sources, such as CBS and Amazon Prime. Syndicated advertising comes to individual stations/outlets in connection with contracted programming, such as *The Ellen DeGeneres Show* or *Wheel of Fortune*. As is the case with radio, the media planner and buyer decide which television outlets to use, although in programmatic advertising (page 47), placement and purchasing are managed automatically through software and algorithms: Machines negotiate with machines for ad prices and placements. For a television commercial, a strategic writer crafts a message based on a predetermined strategy and designed to reach a particular target audience (see Strategic Message Planners and Creative Briefs, pages 171–195).

Even before the advent of online and wireless media, the television audience was becoming increasingly segmented by the growth of cable television. The success of streaming services such as Netflix has further divided the audience. Because it has become more difficult to reach a large audience with just one commercial, successful television commercials target well-defined segments of consumers. Unlike specifically formatted radio stations, where advertisers usually purchase placement in certain day parts, such as morning or afternoon "drive time," television advertisers purchase time in specific programs based on the demographic, psychographic and behavioral nature of the audience each program attracts. And, again, video ads also are posted on company websites and in social media such as YouTube and Facebook.

Video is the most powerful and widely used of the advertising media. Its integration of sound, pictures and motion leaves strong impressions on viewers. Like its cousin audio, video is a linear medium with fleeting messages. However, video requires more of an audience's attention than does audio—a good thing for advertisers but a bad thing if you are driving a car. Thanks to the four screens, the reach of a video ad is unsurpassed. However, it is also the most expensive medium when it comes to production and placement.

**KEY TO SUCCESS**

An effective video ad stays on strategy. It uses both sight and sound to rise above the clutter of media messages. It grabs and holds the attention of viewers.

## Format/Design

Because of the considerable investment of time and money in video ad production, several steps usually precede production. As is the case with all advertising, the first step involves developing the strategic message. Beyond that, producers often write a treatment, a detailed narrative of what the commercial will look like, where and how it will be photographed and other logistical/technical requirements. Scripting is the next step. Television commercial lengths vary, but the most common lengths are 10, 30 and 60 seconds. Depending on how the client and/or producer wishes to proceed, the script can be one of two—and sometimes both—forms.

### *Written Script*

Some similarities exist between this script format and the ones used for audio production. Both provide instructions for bringing together the various production elements. Both include traffic information (see page 204) at the top to ensure that the paperwork goes when and where it is needed. Both often use a two-column format. However, here the similarities end. Because of video's added visual dimension, the left column of a two-column video script contains visual information. This includes descriptions, lengths, widths and sometimes angles of camera shots, as well as descriptions of any special effects or graphic information. The right column details audio information. Write the spoken passages in a large, easy-to-read typeface, and double-space the lines. Because of the two-column format, you often will need more than one page. Number and label each page. Do not divide words or key phrases on different lines of copy or on different pages. And, as is the case with audio advertising scripts, provide a pronouncer for difficult words and names.

### *Storyboard*

A storyboard is a more visual demonstration to a client or an account executive of how the words and pictures will be married into a single persuasive message. In a storyboard, drawings for each camera shot depict the sequence of action, camera angles, widths, settings, special effects and graphics to be used. Depending on the writer's skills, these drawings may be as simple as sketches using stick figures. However, in some large agencies, storyboards often appear to be works of art. Accompanying each drawing is the relevant audio information. The obvious advantage of the storyboard approach is that it forces the writer to think visually when creating the advertisement.

## Content and Organization

In terms of content and organization, practically everything discussed in the audio advertising section (pages 203–209) holds true for video. But video has the added dimension of visual communication. To suggest that it is audio advertising with pictures is a poor use of the medium. While a goal of an integrated advertising campaign (pages 42–43) is to have messages in different media that reinforce one another, that doesn't mean you *have* to use exactly the same words in all media. With audio, words carry most of the burden of delivering the message. In video, pictures and graphics share that burden.

In writing your script, be sure to follow the "Tips for Visual Storytelling" on pages 32–34—particularly the passage on so-called "hit-and-run writing." Marry images and words so that the two complement each other.

In the left column of your script, describe each shot concisely but thoroughly. A shot is a camera placement. When the camera physically moves to a different location, a new shot begins. For each shot, specify the length (how many seconds the shot will last) and the width (wide shot, medium shot or close-up). With storyboards, you also can indicate the angle of each shot.

A series of shots that convey a single action within the same setting is called a sequence. If you think of a single shot as a sentence, then think of a sequence as the video equivalent of a paragraph. Just like a paragraph in print, each sequence has a topic sentence, which is the establishing shot—usually a wide shot that sets the scene. The establishing shot provides context for shots that follow.

To carry our paragraph analogy further, good writers like to vary sentence length to avoid boring, predictable patterns. They want to maintain their readers' interest. This is also true in the world of television and film. For each new shot within a sequence, change the width, the length and the angle. For example, if we have a six-second medium shot of a woman at a desk, our next shot might be a three-second close-up of her face. The angle of our first shot might be straight ahead. The angle of our second shot, for variety and movement, might be about 45 degrees to the left. Remember that video is an active medium: Viewers expect movement and variety.

The terms *wide, medium* and *close-up* are relative, flexible terms. Commonly, a wide shot might show a room with people in it. Again, wide shots are the most common form of establishing shots. A related medium shot could be two people talking, shown from their waists up. A related close-up could be a person's face. However, in a commercial for contact lenses, a wide shot might be a man shown from the waist up. A related medium shot could be his face, and a related close-up could be his eyes.

Generally, begin your script with a wide shot—an establishing shot. Avoid putting two wide shots or two medium shots back-to-back. The lack of detail and variety becomes boring. Again, remember that viewers expect movement and variety. Close-up shots can follow one another, especially if the visual element changes. For example, we might go from a close-up of a woman's face to a close-up of her hand tapping a desktop.

In a standard 30-second commercial, shots generally range from two seconds to eight seconds in length. Again, variety is important. A series of five-second shots will bore viewers.

## VIDEO ADVERTISEMENT TIPS

1. **Review audio**: Remember the tips for audio advertisements. All of the guidelines discussed for audio on pages 203–209 hold true for writing video advertisements.
2. **Think visually**: But also remember that video is not audio with pictures. Work toward marrying pictures, graphics, words and music into a cohesive and effective message.
3. **Tell a story**: Tell a story that relates to both the reality and dreams of the target audience. In successful commercials, viewers project themselves into the situations being portrayed. That means that the advertisement must be in touch with reality—either the reality of the audience's current situation or that to which it aspires. Review the different creative approaches on page 205.
4. **Stay on message**: Entertainment is a strategy, not a goal. It's fine to use entertainment to raise a commercial above the clutter of competing media messages. But if the viewer doesn't remember the purpose for the commercial—the strategic message about the client's product—then the ad is a waste of money.
5. **Test your message**: This is good advice for any advertising message. But it's especially true for video advertising because of its cost. It's a lot easier to fine-tune the message during the preproduction stage than it is after the ad is—using the jargon of the business—"in the can." Large agencies and companies will first produce several versions of a commercial and conduct private audience tests before public release.
6. **Respect the audience**: Remember that the viewers have most of the power in this relationship. If they don't like you or think you don't like them, the ad will be a wasted effort. Earn viewers' respect by talking to—not at—them. And because of the size of video audiences, many who see your message may be outside your target audience. They deserve—and will demand—equal respect.

# Video Advertisement Production Script

Title: Riverview Image Ad
Client/Sponsor: Riverview Community Medical Center
Length: 60 seconds
Air Dates: June 1–September 1

| | |
|---|---|
| WS—Riverview Community Medical Center front entrance exterior (:07) | MUSIC:<br>(Light, bouncy mood music. Establish, then under)<br><br>ANNOUNCER:<br>For more than 60 years, Riverview Community Medical Center has been your hometown health care provider. |
| MS—Doctor and nurse looking at a child's tonsils with Mom watching (:05) | Hometown professionals. |
| CU—Mom's face, smiling at the scene (:03) | The best care. |
| MS—The doctor handing the child a lollipop (:02) | Neighbor to neighbor. |
| WS—Family members visiting a new mom in her room (:05) | We have been there . . . |
| MS—Nurse comes into the room with a newborn baby and hands the infant to Mom with Dad watching (:04) | . . . in good times and bad . . . |
| CU—Mom, Dad and baby (:03) | . . . from generation to generation. |
| WS—Entrance to Rehab & Sports Therapy Services (:05) | For the health of our community, we have added two new services. |
| MS—Therapist working with a patient (:06) | Rehab and Sports Therapy . . . and the Center for Joint Replacement. |
| CU—Therapist and patient at work (:06) | Just two more examples of how we meet the needs of the people of Gleason County. |
| MS—Small congregation of a cross-section of RCMC staff, doctors and nurses (:05) | Riverview Community Medical Center. . . . |
| CHYRON: (centered in the lower-third of the frame):<br>Riverview Community Medical Center / 555-1983 | MUSIC:<br>(Up and then fade, out at :59) |
| FADE TO BLACK AT :59<br><br>RT -- :59 | |

###

## Video Advertisement Storyboard

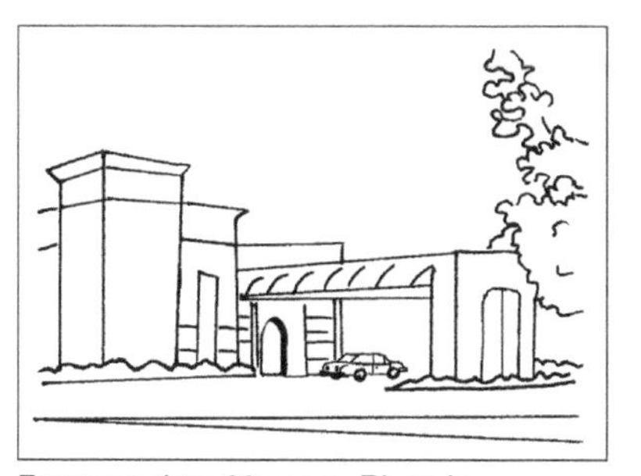

For more than 60 years, Riverview Community Medical Center has been your hometown health care provider.

Hometown professionals.

The best care.

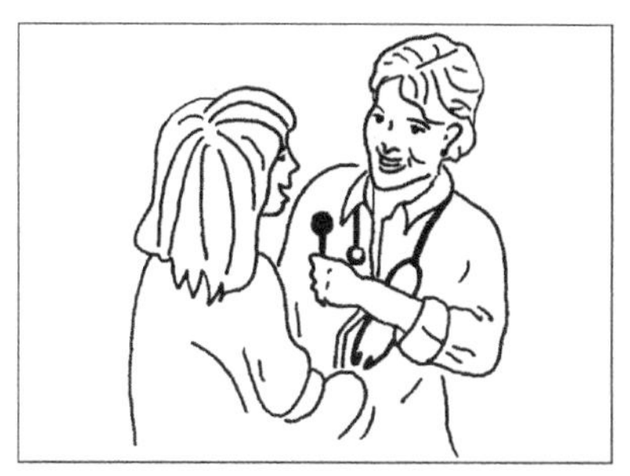

Neighbor to neighbor.

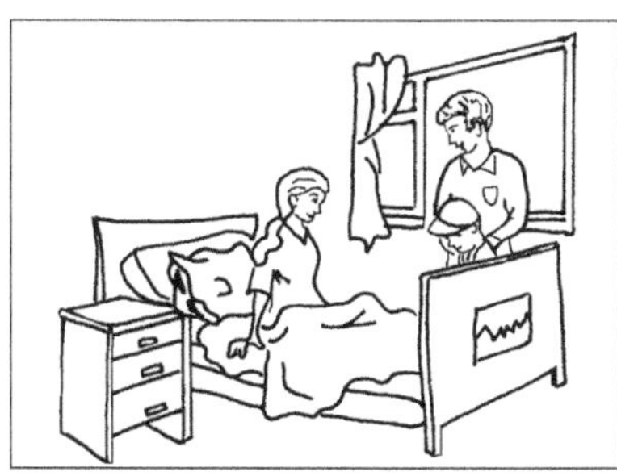

We have been there . . .

. . . in good times and bad . . .

. . . from generation to generation.

For the health of our community, we have added two new services.

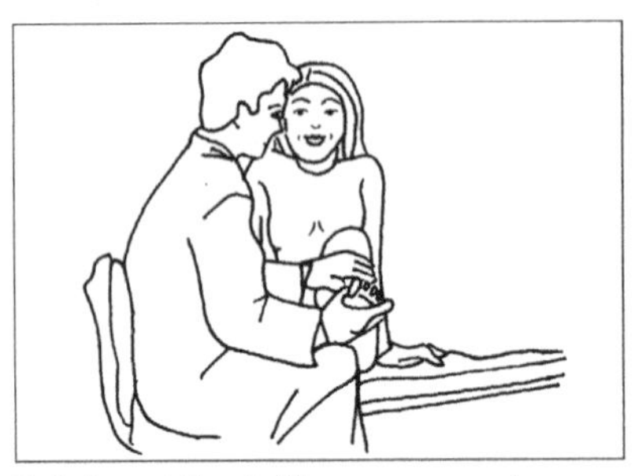

Rehab and Sports Therapy . . . and the Center for Joint Replacement.

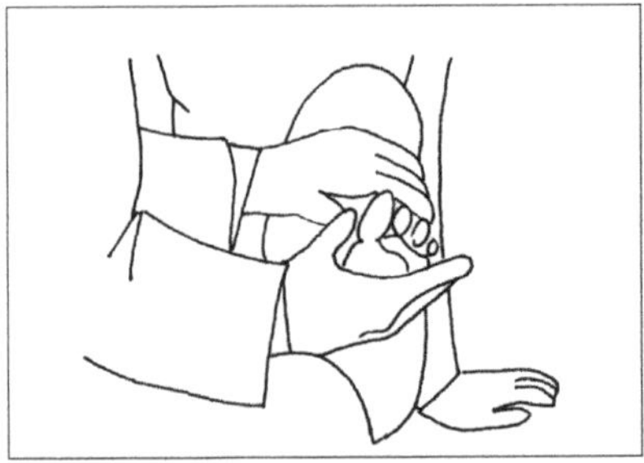

Just two more examples of how we meet the needs of the people of Gleason County.

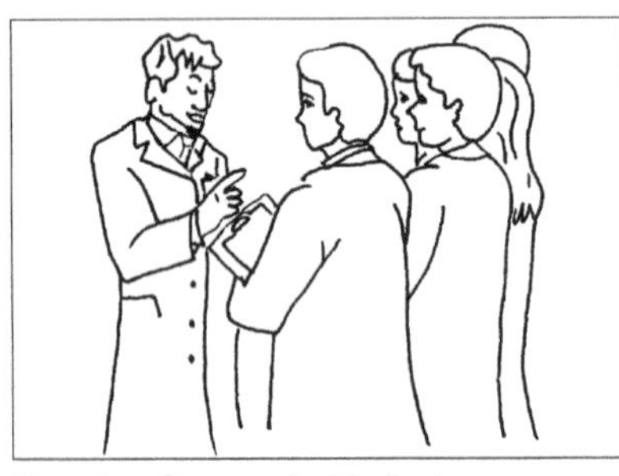

Riverview Community Medical Center. . . .

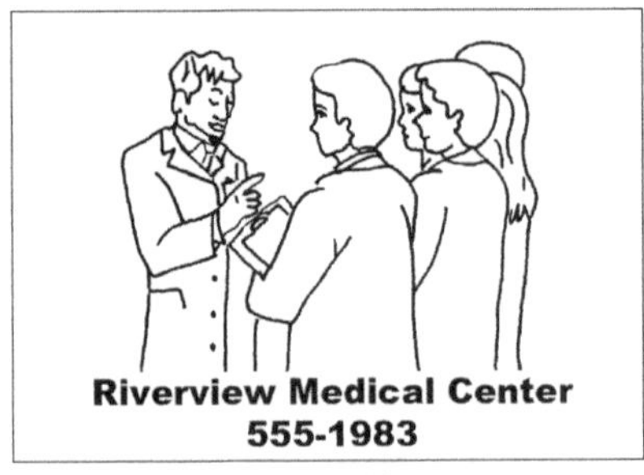


. . . your hometown health care team.

# 3H Digital Advertisements

## Purpose, Audience and Media

Digital advertisements are persuasive messages that identifiable sponsors pay to deliver via phones, tablets, laptops and similar technologies with screens. Unlike in print and broadcast ads, the sponsor may not be identified in the visible ad because digital ads often link to a website, generally called a landing page, that may contain the first identification of the sponsor.

Digital ads can be highly targeted to specific audiences. When we go online, particularly when we're browsing, we leave a history—often via bits of data called cookies—that store and share data about the sites we visit and the searches we conduct. Cookies and related technologies allow websites such as Facebook, Amazon and CNN to present digital ads based on our online history. Additionally, ad-placement technologies known as programmatic advertising use deep data and algorithms to automatically place ads for an advertiser, ideally zeroing in on the most appropriate media for reaching the desired consumers (see page 47). This means, of course, that understanding our target audience's demographics, psychographics and behaviors has never been more important.

Members of your target audience rarely go to their phones, tablets or laptops to read ads. (Google searches can be an exception.) Thus, a good digital ad must be concise and attract attention without being annoying. For more information about online target audiences, see Writing for the Web (pages 19–25) and Websites (pages 84–93).

The medium for a digital ad is, generally, a website, social media platform or app displayed on a phone, a tablet or a laptop. However, because such ads are digital, they can easily migrate beyond their original placement though forwarding, sharing, retweeting and related technologies.

**KEY TO SUCCESS**

Digital ads often must persuade the viewer to click or tap through to an appropriate landing page. This can be done by using strategies such as listing benefits or creating an irresistible sense of mystery.

## Format/Design

The physical appearance of digital ads, including restrictive character counts and image/video size restrictions, may be the most diverse and rapidly changing area of strategic communication. Marketo, which is part of Adobe's marketing services operations, divides the world of digital ads into the following categories.

### *Pay-Per-Click (PPC) Ads*

These are the familiar search-related ads, often text-based, that appear at the top of Google or other search-engine searches. Advertisers pay a premium to appear at the top of the list, and they pay for each time a viewer clicks through to a landing page.

### *Display Ads*

Display ads generally include images. They come in five varieties: static ads, video ads, cinemagraphic ads, Flash ads and pop-up ads.

#### *Static Ads*

Marketo calls static ads "the bread and butter" of digital advertising. In other words, static ads are basic and important. A static ad contains at least one visual element and, usually, a small number of support words (see page 223 for an example). They often contain a call-to-action button that can lead viewers to a landing page. Narrow horizontal static ads often appear at the top of a webpage. Square or narrow vertical static ads often appear on the right side of a webpage, sometimes called the right rail.

#### *Video Ads*

As the name suggests, video ads feature a video that often plays automatically, with the audio muted, within a rectangular or square ad space. Viewers can opt to turn up the audio. Video ads often feature a call-to-action button that appears either at the end of the video or as soon as specific benefits are presented.

#### *Cinemagraphic Ads*

Cinemagraphic ads combine static images with isolated movement: At least one element of the image moves, ideally drawing the viewer's attention. For example, hair might flutter, text might pop in and out or lightning might flash in the image's background. Other visual elements of the ad remain unchanged.

### *Flash Ads*

Often created with the Adobe Flash program, Flash ads can combine sound, video and animation.

### *Pop-Up Ads*

These often unpopular, intrusive ads can be highly effective if they quickly and clearly address viewer interests. Pop-up ads, generally squares or rectangles, overlay a portion of the screen that the viewer is trying to see.

## *Mobile Ads*

Most display ad formats can function as mobile (phone) ads, but Marketo cautions that success depends on what function the phone is using. For example, cinemagraphic ads can work in a phone's browser but often not in apps or texts. Video ads can work in browsers and apps but not in SMS texts (see page 261). As phone technology improves, mobile ads will transfer more easily from function to function—from browser ads to text ads, for example.

## *Social Ads*

Marketo identifies three categories of social ads:

1. Promoted posts that appear in regular news streams in platforms such as Instagram, Facebook and Twitter. They appear as regular posts and often include a label such as "Sponsored" or "Promoted."
2. Display ads that often appear on the right of the screen, particularly in laptops and tablets. LinkedIn includes such ads.
3. In-stream video ads in platforms such as YouTube. Ads in those videos can appear as an introductory video ad or as an in-stream insertion several minutes into the video the viewer was watching. Ads within videos also can appear as display ad overlays, similar to pop-ups, generally at the bottom of the video.

An increasing number of studies indicate the power of multimedia ads, particularly those with video.

Be sure to maintain a consistent tone and appearance from your digital ad to its landing page. The landing page should be a logical extension of the digital ad, whether the purpose is to make a sale, identify a customer for possible future sales or something else altogether. The purpose of the digital ad–landing page combination is to "convert," whether that means generating sales or sales leads, promoting phone calls, increasing signups or initiating online chats with potential customers.

For example, in the ad–landing page combination on pages 223–224, note these points:

- The offer is exactly the same on both the ad and its landing page.
- Both components use the same wording to describe the offer.
- The color scheme is the same, so it's easy for visitors to tell they're in the right place.
- The *flavours!* logo in the lower right isn't linked to the homepage so viewers will stay on this page. Landing pages generally don't link to other pages. Your goal is to keep the viewer on this page until, ideally, they convert.
- Similarly, no header or footer links are present, nor are there any social media links that could take viewers away from this page.
- The page consistently promotes a free offer: The headline, copy, images and call-to-action button are relevant to the offer.
- The orange color for the call-to-action button contrasts with the rest of the page, drawing attention to itself.
- "Click for Coupon NOW!" lets viewers know exactly what clicking the button does—gets the coupons sent to them.

And there's actually one more tactic after the digital ad and its landing page: Don't forget a thank-you page. Connect your landing page to an optimized thank-you page, which allows you to acknowledge and thank the visitor for converting, making them feel appreciated and building a stronger relationship with them. Most important, it can present you with another conversion opportunity. A thank-you page can include:

- A sincere thank-you note to the customer
- An image of the offer, if applicable
- An explanation of the next step (how they will redeem the offer)
- Any related offers

Fortunately for strategic writers, this incredible variety within format and design begins with a written document. In addition to an SMP or a creative brief, that document can be a digital or paper sketch, a brief written description or, in the case of video ads, a short script similar to a television advertisement script (page 214).

Ad copywriters often work with designers. For more information on digital designs and formats, see Websites (pages 84–93). Digital ad designers should be aware of the particular technical specifications for each platform. Don't exceed specified size requirements for photos or videos. Twitter ads, for example, differ from Facebook ads. Digital ads should be "optimized" for the technical abilities of each specific digital platform. Mobile ads, for example, should be optimized for mobile delivery.

## Content and Organization

The impressive and growing diversity of digital ad formats makes it difficult to specify standard content and organization. The writing within a display ad, for example, will differ from the writing within a pay-per-click ad. But besides the images, photographs or other graphics included within most digital ads, most have one or more of the following written elements:

- Headline—generally followed by only one sentence or sentence fragment
- Product name—can stand alone or appear with an image and other text elements
- Slogan—can stand alone or appear with an image and other text elements
- Benefits—are concise, can appear sequentially and often precede a product name or company name
- Question(s)—can stand alone or appear sequentially, one after another, ideally urging the viewer to click or tap through to a landing page
- Call to action—is a blunt request for the viewer to do something specific
- Request to click or tap (a form of call to action)—can be implied (most of us know to click or tap) or can be explicit, as in "Find Out Now" or "Learn More" or even "Tap Here," all printed on a button
- Logo—rarely stands alone, generally appearing with other written elements, such as a slogan or headline, or images, such as a product shot

Studying the ads you receive via your phone and other screens can provide examples of each of these text elements.

A written element that almost never appears as part of a digital ad is a paragraph of two or more sentences. Digital ads are concise, many containing fewer words than a haiku, the Japanese poem that has three lines and 17 syllables. In digital ads, a written benefit, for example, is rarely more than four words. In-stream video ads are rarely more than 30 seconds.

Because benefits must be concise, the "16 words that sell" on page 200 can be particularly useful. Effective digital ads quickly deliver what is a called a "value proposition" to viewers, swiftly showing them how a particular product could improve their lives.

The amount and kind of writing in digital ads depends on the format. A display ad may be as simple as a slogan and a product image. Pop-up ads tend to be annoying and so must immediately convey a powerful benefit to the viewer. Landing pages can monitor the number of clickthroughs, thus supplying data that might help you identify which digital ad format appeals to your target audience.

## WEB/MOBILE AD TIPS

1. **Be strategic**: Complete a strategic message planner (pages 171–191) before you begin to write a digital ad.
2. **Imitate**: Digital ads probably bombard you via your phone and social media such as Twitter and Facebook. Study them. Which ones appeal to you? Why?
3. **Simplify**: Don't underestimate brevity and minimalism—which may sound odd, given the technological possibilities of digital ads. Almost all of us have been irritated by floating pop-ups with excessive animation.
4. **Diversify and compare**: If your budget allows, consider using slightly different versions of the same ad to see which earns the most clickthroughs.
5. **Optimize**: Be sure to optimize landing page websites for mobile devices. Consumers now often link to the site via phones, not laptops.
6. **Follow instructions**: Google, Facebook, Twitter and other digital advertising platforms have easily accessible templates and guidelines to help you create effective ads. Pay particular attention to character-count and image restrictions.

## Digital Static Ad

Credit: GLandStudio.shutterstock.com/ Nitr.shutterstock.com/ ria_airborne.shutterstock.com

## Landing Page

**Get 2**

**FREE Meals**

**with proof of purchase of 5 *flavours!* entrées**

**Tell us about yourself, text 5 UPC codes and we'll send you a coupon for 2 FREE entrées!**

| First Name | Last Name |
|---|---|
| Email Address | Zip Code |

We respect your privacy.

- 34 entrées to choose from
- Each less than 400 calories
- Organic ingredients
- Created by award-winning chefs

Convenient. Delicious. Simple. Healthy.
*flavours!* A perfect balance for life.

Credit: GLandStudio.shutterstock.com/ Nitr.shutterstock.com/ ria_airborne.shutterstock.com

# 31 Radio and TV Promotions

## Purpose, Audience and Media

The most effective and efficient means for radio and television outlets to promote programming and establish brand image is on-air promotional announcements using their own stations or channels. These announcements—commonly referred to as *promos*—are often tied to advertising, public relations and promotional efforts in other media, especially social media. Promos are designed to build and maintain a desired audience. Strategic writers can design these announcements either to attract an audience to a particular program or to nurture the relationship between the audience and the media outlet. The latter is particularly true for broadcast (as opposed to cable) stations, which are required to demonstrate that they operate in the public interest as part of government licensing requirements.

The purpose of a promo dictates its target audience. Program promotions target audiences of specific programs. For example, fishing shows tend to attract a predominantly male audience, whereas soap opera audiences are predominantly female (though it can be dangerous to generalize in discussing target audiences). Image promotions, however, help create a desired brand image, or position, for an entire station, channel or network, which generally has a more diverse audience. Highlighting a media outlet's competitive edge over its competition is the key to successful image promotion. Image promos also help attract viewers and listeners to programming geared to a more diverse audience, such as the evening news.

Broadcast and cable promotion became a significant element of programming strategy starting in the 1970s because of a combination of factors. The Federal Communications Commission required national television networks to open the first hour of prime viewing time to local programming. Hollywood and programming syndicates seized the opportunity by providing original and previously broadcast programs for local use. Radio and TV promotion is particularly important in the Digital Age because technological advances have made it easier to produce and distribute programming. Today's focus is on cross-platform promotion (also known as *cross-promotion*), the marriage of messages across traditional, online and social media. The idea is simple: Expose people to the same promotional message in as many venues as possible. The more people you attract and the stronger the relationship with the audience you build, the higher the advertising rates you can charge.

Although this book focuses on the creation of messages, it is important to remember that radio and television promotions also involve a wide range of tactics, including special events, such as concerts or contests. All promotional messages, however, must complement the overall strategic communications plan.

**KEY TO SUCCESS**

Through crisp writing and effective repetition, a successful promo delivers an unambiguous message, often about program name, time and station. It closes with a clear call to action.

## Format/Design

On-air broadcast/cable promotions follow the same script formats used in audio advertising (pages 203–209) and video advertising (pages 210–216). This should not be surprising, since the only major difference between an advertisement and promotional announcement is the client.

## Content and Organization

Promos follow the same preliminary research and planning process used in advertising. In other words, complete a strategic message planner or creative brief (pages 171–195) for your promo. This book's sections on audio and video advertising also include advice that can apply to successful promos.

Besides basic image promos, there are two kinds of on-air broadcast/cable promos: topical and generic. A *topical* promo provides information about a specific episode airing at a particular time. Examples include a promo teasing the story line of an upcoming episode in an entertainment program. Just like that last piece of pizza in your refrigerator, topical promos have a limited shelf life. Once the particular episode featured in a topical promo has aired, the promo is of no further use.

Contrast topical promos with generic promos: A *generic* promo is designed for broadcast at any time. Rather than promoting a specific episode of a program, a generic promo reminds viewers and listeners of when and where a specific program airs. Generic promos often are referred to as *evergreen* promos because they are always fresh and never out of date. An example is the promo that reminds viewers to "Watch Eyewitness Action News (insert TV channel here) every night at six." That promo is good as long as Eyewitness Action News Whatever sticks to that same program schedule.

As we've mentioned, repetition of key information is important in promo writing. (Notice how we cleverly used repetition to reinforce this point!) In radio and television, your message competes with a variety of distractions for the audience's attention. Remember to include the key information the audience needs to know:

- The name of the program
- The time and day it is broadcast
- The station/channel/network on which it is broadcast. Don't assume that just because people watch a program, they know which station they are watching. Through channel surfing, they may be a part of your audience by chance rather than choice. The goal of a promo is to have viewers and listeners remember your station when they make a programming choice—*and* when they fill out ratings surveys.

It is not enough to tell viewers or listeners the *when* and *where* of your program just once—they may have missed it. During a 30-second promo, consider repeating key information three times—in the opening sentence of the promo, once in the middle and at the very end for closing emphasis. This is easier to do in television, where visual images complement the words.

In writing your promo scripts, be sure to follow the guidelines for Writing for the Ears on pages 29–34.

## RADIO/TV PROMO TIPS

1. **Think before you write**: Is the promo image generic or topical? Have a clear understanding of this before you do anything else.
2. **TDT—Tease; don't tell**: You want to pique interest, not give the plot away. Who would want to watch a murder mystery if the promo announced "who done it"?
3. **Transition, transition, transition**: In topical promos that use sound or video clips, the announcer should create smooth transitions from clip to clip.
4. **Don't promise what you can't deliver**: Remember that the goal is a relationship, not just a one-time audience. If you make a promise—such as a special appearance by a popular celebrity—but don't deliver on it, you will lose credibility. Hyperbole is fine. But if you go too far, you will lose the audience.
5. **Be conversational**: Take some liberty with the language. Write the way people talk. Sentence fragments and questions are fine. You also can stretch the rules of grammar—just as long as doing so makes sense to the audience and doesn't detract from the message.
6. **Target**: Direct the promo to a specific audience with clearly defined demographic, psychographic and behavioral characteristics.

*continued*

## RADIO/TV PROMO TIPS

7. **Repeat yourself**: Did we mention this one? Identify the program, time and station more than once. Be sure to close with this information and a call to action.
8. **Write light, tight and bright**: In the clutter of today's media, the goal of any promo is to be remembered. Your listeners and viewers value creativity—as long as it doesn't come at the expense of giving them the key information. Get the audience's attention and keep it.
9. **Review**: Follow the tips presented for audio and video advertising—they apply to promos as well.

## Announcer Continuity Radio Promotion

**Title:** Health Beat

**Client/Sponsor:** Station promo

**Length:** 20 seconds

**Air Dates:** Until further notice

Want the latest news in health and medicine? Join Riverview Radio weekday mornings at 7-15 for Health Beat with Doctor Marco Podestra (pah-DESS-trah). Weight loss tips. The warning signs of cancer. Skin care. We cover it all. Health Beat is brought to you as a public service of this station and the Riverview Community Medical Center. That's Health Beat with Doctor Marco Podestra. Weekday mornings at 7-15 on Riverview Radio.

# # #

## Radio Promotion Production Script

Title: Don't Skip a Beat
Client/Sponsor: Station Promo
Length: 30 seconds
Air Dates: November 8–14

| | |
|---|---|
| SFX: Sound of a human heartbeat, with the rate gradually increasing. (Establish, then under) | |
| ANNOUNCER: | 60 times a minute. 36-hundred times an hour. 86-thousand times a day. More than 30-million times a year. It is easy to take your heart for granted. . . . |
| SFX: Heartbeat stops. Flatline tone replaces heartbeats. (Establish, then fade) | |
| ANNOUNCER: | . . . until it stops. |
| MUSIC: Action News Six Theme. (Establish, then under) | |
| ANNOUNCER: | This week on Action News Six at Six. Join medical beat reporter Melissa Cochran for her special health series "Don't Skip a Beat." |
| ANNOUNCER: | Melissa will bring you the information you need to keep your loved ones safe. And we will have heart specialists from Riverview Community Medical Center standing by to answer your questions. |
| MUSIC: (Fade)<br>SFX: (Re-establish sound of heartbeat) | |
| MELISSA COCHRAN: | I'm Melissa Cochran. Join me every night this week for life-saving information. That's "Don't Skip a Beat." Every night this week on Action News Six. |
| SFX: (Heartbeat stops abruptly) | |
| MELISSA COCHRAN: | This is one series you can't afford to skip! |

###

## Television Promotion Production Script

Title: 11:00 News Promo
Client/Sponsor: Station Promo
Length: 29 seconds
Air Dates: Until 11 p.m., Tuesday

| | |
|---|---|
| ESTABLISHING SHOT: Paul Rodgers, evening news anchor, at news set. | MUSIC:<br>News theme. (Establish, then under)<br><br>PAUL RODGERS:<br><br>Tonight at 11 on Action News Six.... |
| VO/SOT: Pictures of police working a crime scene. | PAUL RODGERS:<br><br>Just how safe is it to live in Riverview? Our Michelle Masters has the latest crime statistics. Get them in her special report "Mean Streets." |
| REPORTER STAND-UP: Medical beat reporter Melissa Cochran in front of the emergency room at Riverview Community Medical Center. | MELISSA COCHRAN:<br><br>I'm Action News Six Medical Beat Reporter Melissa Cochran. Did you know that one in four Americans suffers from heart-related illness? What do you need to know to protect your family? Watch my special report "Don't Skip a Beat." |
| MS—Rodgers at news desk | PAUL RODGERS:<br><br>Biff has the latest in college hoops. Bev has your weekend forecast. That's tonight at 11 on Action News Six! |
| FADE TO BLACK AT :29 | MUSIC: (Fade at :29) |

###

# 3J Public Service Announcements

## Purpose, Audience and Media

Audio and video public service announcements are persuasive messages carried without charge by radio and television outlets on behalf of nonprofit and social-cause organizations. Broadcast stations carry PSAs to fulfill federal licensing requirements that those outlets serve the public interest. Cable TV and other media outlets also carry PSAs to demonstrate their commitment to their audience.

PSAs also can resemble magazine advertisements, suitable for printing or posting, but in this section, we'll focus on audio and video PSAs.

PSAs look and sound like commercial announcements. They are targeted communications that contain a call to action. However, two major differences separate PSAs from commercials. The first has to do with control. Commercial announcements use controlled media. Advertisers pay for the right to choose the form, timing and placement of messages. However, PSAs use uncontrolled media. With PSAs, media outlets—and not the message provider—make decisions about whether and when to use a PSA. The second major difference is that commercial announcements tend to promote marketplace transactions, whereas PSAs more often promote social causes and behavioral change. Paid commercial announcements promoting social causes and behavior are not PSAs. Neither are network or station promotional announcements designed to sound like PSAs (such as NBCUniversal's long-running "The More You Know" campaign).

Like all persuasive messages, PSAs target specific audiences. However, because they are not guaranteed airtime, they also target the media preferred by the desired audience. For example, it would not make sense for the Future Farmers of America to send a PSA to an urban radio station that plays hip-hop music. It seems unlikely that the FFA's desired audience listens to that station. A better fit might be rural country music stations—though, as we've noted before, we should be cautious about predicting any target audience's characteristics without doing the research. Similarly, organizations often create PSAs with a station's format in mind, such as a PSA featuring a hip-hop music artist for broadcast exclusively on hip-hop music stations. Conversely, television stations affiliated with one network may not wish to show PSAs featuring stars from another network. Local media outlets are more likely to carry messages of local interest, and national media outlets are more likely to carry messages that are relevant to a national audience. Although the CBS television network is not likely to broadcast a PSA for a local blood drive, the local CBS affiliate might.

In the United States, the federal government licenses broadcast stations, and, as a part of the licensing agreement, those stations must serve the public interest. Although cable television outlets do not face the same requirement, they may have to carry PSAs as part of a local franchising agreement. A media outlet chooses which PSAs to broadcast based on its strategic and logistical needs. It will select messages compatible with the audience it wants to attract. Conversely, it will shy away from controversial material that may alienate an audience. Because PSAs do not directly add to the media's bottom line, those that require the least amount of pre-broadcast preparation are most likely to be accepted.

With that in mind, it is not unusual for organizations seeking public service airtime to provide a package of PSA materials that give media outlets maximum flexibility, especially in terms of message length. These packages typically include 10-, 15-, 20-, 30- and 60-second versions of the same announcement.

How should you distribute your PSAs? Williams Whittle, an agency that specializes in PSAs, recommends using a distribution service, such as Cision, to distribute a national PSA, with additional information that can lead recipients to your organization's digital newsroom for more options. The favored lengths for audio and video PSAs, Williams Whittle reports, are 15 and 30 seconds.

Just as with video news releases, today's online and social media allow organizations to bypass mass circulation media and direct their PSAs to a specified audience. It is commonplace for organizations to feature PSAs in their websites, YouTube channels and digital newsrooms.

**KEY TO SUCCESS**

It's not enough to consider the needs of the target audience. You must also consider the needs of the media outlet through which you hope to reach that audience.

## Format/Design

Public service announcements follow the same script formats used in audio advertising (pages 203–209) and video advertising (pages 210–216). This should not be surprising because of the aforementioned similarity to commercial announcements.

## Content and Organization

PSAs follow the same preliminary research and planning process used for advertising. Yes, for the one-millionth time, we're telling you to complete a strategic message planner or creative brief (pages 171–195) before you begin to write a persuasive message.

In terms of content and organization, virtually everything discussed for audio advertising (pages 205–209) and video advertising (pages 210–216) holds true for PSAs. Because of the nature of the persuasive messages delivered by PSAs—often for social causes or to create changes in behavior—it's especially important for a PSA to reflect the values of the sponsoring organization. Advocates of particular perspectives are often held to a high standard of conduct.

A digital newsroom PSA package can contain the following:

- **A cover letter or message**. This is written like a one-page sales letter (see pages 265–272). Your challenge is to show why this message is relevant to a media outlet's audience and/or its strategic interests.
- **A list of materials**. You may have only a few moments to win over a program director or public service director. This list makes their job easier, and, therefore, they may be more likely to spend time reviewing the package.
- **Scripts of recorded messages and/or announcer continuities** (page 204). Include these to ease the work of the program director or public service director. Williams Whittle reports that announcer continuities are increasingly popular for radio because they allow well-known station personalities to participate in the worthy cause, thus boosting the station's reputation.
- **Recorded messages in a variety of versions and lengths**. People like having options. Program directors and public service directors are no different.

Be sure to follow the guidelines for Writing for the Ears on pages 29–34.

## PSA TIPS

1. **Beggars can't be choosers**: Offer media outlets materials that do not require pre-broadcast preparation. The easier you make it for the broadcaster to use your PSA, the more likely it is to be accepted. Multiple versions of the message increase a media outlet's options.
2. **Remember each media outlet's audience**: Its audience should be the same as your desired audience. Media target programming to attract specific audiences. When seeking public service airtime, identify the media that reach the desired audience.
3. **Don't expect a free lunch**: Because media outlets are in the business of making money, they rarely accept PSAs for which space or time has been purchased in other media.
4. **Don't expect premium placements**: Remember that you can't always get what you want. Because PSAs use uncontrolled media, don't expect to see or hear them during prime programming hours. The media charge the highest advertising

## PSA TIPS

rates for the times when they have the largest audiences. It is not likely they will give this time away. (If you're wondering about PSAs played during the Super Bowl, the National Football League requires their broadcast as part of its contractual agreement with the networks.)

5. **Don't pinch pennies**: Although media outlets do supply airtime without cost, don't assume that PSAs are inexpensive. PSAs often cost as much to produce as any commercial announcement. This shouldn't be surprising: PSAs, like commercial announcements, need to rise above the clutter to gain and maintain an audience's attention.
6. **Remember who you are**: PSAs must reflect an organization's values and mission. Typically, nonprofit and social organizations use PSAs as a tactic in their strategic communications. These organizations are especially sensitive to the opinions of key stakeholders. Therefore, all messages delivered on their behalf are viewed with a critical eye.
7. **Review**: Have one more look at Writing for the Ears, pages 29–34.

## Audio PSA Production Script

Title: Project Graduation
Client/Sponsor: Riverview PTA
Length: 30 seconds
Air Dates: May 20–June 4

| | |
|---|---|
| MUSIC: "Pomp and Circumstance." (Establish, then under) | |
| ANNOUNCER: | You've finally made it. The big day has arrived. You've graduated! It's time to celebrate. |
| SFX: Pop-top on a can of beer | |
| ANNOUNCER: | But the *way* you celebrate may affect *how long* you celebrate. |
| SFX: Turning key on a car ignition | |
| ANNOUNCER: | As you celebrate graduation, please remember that drinking and driving don't mix. |
| MUSIC: (Ends abruptly) | |
| SFX: Car crash sounds | |
| ANNOUNCER (after short pause): | Don't make this a day that your family and friends will remember for all the wrong reasons. Celebrate safely. A reminder from the Riverview P-T-A. |

###

# Video PSA Production Script

Title: Take a Hike
Client/Sponsor: Riverview Medical Association
Length: 30 seconds
Air Dates: Until 9:00 a.m., September 15

| | |
|---|---|
| VO—Scenes from last year's cancer walk (:06) | MUSIC:<br>"Hit the Road, Jack" by Ray Charles. (Establish opening chorus, cut to instrumental music, then under) |
| CHYRON—Tell Cancer to Take a Hike/9:00 a.m. Saturday, September 15 (:08) | ANNOUNCER:<br>Join your friends and neighbors on Saturday, September 15th, as they tell cancer to "take a hike." You can run, walk or ride the short course while raising money for cancer research. |
| CHYRON—555-0000 (:07) | ANNOUNCER:<br>If your company or organization would like to participate, call 555-0000 or sign up at the start line in City Park. |
| CHYRON—Riverview Medical Association/www.riverviewmedassoc.org (:08) | ANNOUNCER:<br>Tell cancer to take a hike. Saturday, September 15th, at City Park. Brought to you by the Riverview Medical Association. |
| FADE TO BLACK AT :29 | MUSIC: (Fade at :29) |

###

SECTION

# 4

# Strategic Writing in Sales and Marketing

## OBJECTIVES

**In Section 4: Strategic Writing in Sales and Marketing, you will learn to write these documents:**

- Basic content-marketing tactics
- Social media calendars
- Proposals and marketing communications plans
- Mobile messages
- Sales messages and e-blasts
- Fundraising messages and e-blasts
- Brochures

# 4A Introduction to Sales and Marketing

Marketing is the process of researching, creating, refining and promoting a product—and distributing that product to consumers. Strategic writers participate in every stage of marketing. For example, in the research stage they help develop marketing plans, proposals and business reports. The creating and refining stages include wording on packaging, landing page copy and more reports. The promotion stage usually demands the most writing: proposals, advertisements, news releases, text messages, blogs, podcasts and more. Can overworked strategic writers finally relax during the distribution stage? Of course not. Distribution can require sales-support materials (such as brochures, websites and point-of-purchase displays) and, yes, more reports.

One part of marketing involves directly asking consumers to buy the product. When that's done interactively—face-to-face or through a website or some other medium that allows consumers to respond—it's called sales. Even when the sales process involves a one-on-one meeting, known as personal selling, strategic writing can play an important role: Salespeople often use multimedia presentations, brochures and other written material—product literature—to help explain and promote the product. Strategic writers who help prepare multimedia materials for a sales force work in an area called sales support.

Tension sometimes exists between an organization's sales force and its marketing team. Often, the source of that tension is poor communication. Members of a sales force sometimes believe that the marketing team is out-of-touch with consumers and the realities of trying to sell the product. And the marketing team sometimes believes that the sales force will say or do anything to sell the product. However, an organization should speak with one clear voice when trying to sell a product. In other words, the sales force and the marketing team—plus the product literature, the advertisements, the social media posts and all the marketing communications—should focus on the same strategic message when addressing a target audience. This "one clear voice" philosophy is part of integrated marketing communications, discussed earlier in Section 1 (pages 42–43). Ideally, an integrated, strategic message provides a consistent, beneficial image for a product—and that clear image leads to sales. Communication between the sales force and the marketing team before, during and after a sale is an important part of strategic communication.

Because marketing is product-oriented, it can include advertising and some parts of public relations. And with all those reports, marketing definitely includes parts of business communication. All this inclusion means that the dividing lines among marketing, advertising, public relations and business communication can be blurry. Some documents included in other sections of this book also can be part of sales and marketing. For example, a multimedia news release that announces the launch of a newsworthy product fulfills a marketing function. A good-news business message written to a customer can fulfill a marketing function. The marketing communications plan (pages 252–260), which is the important document that specifies the components of a strategic plan to promote a product, can include tactics from public relations, advertising and other disciplines.

Even professionals and professors sometimes disagree among themselves about the dividing lines that separate marketing, advertising, public relations and business communication. To some, it's all marketing. Others say that because public relations and business communication often focus on groups other than customers, those professions are not entirely part of marketing. For the moment, it's more important for you to focus on the strategic purpose of each document you write. We believe you should focus on a document's goal-oriented reason for existence rather than wondering whether that document is part of marketing or public relations or nuclear physics. A well-trained strategic writer should be ready to tackle any situation that requires the power of good writing to help achieve a goal.

Just as marketing communication involves many disciplines, including public relations and advertising, it involves many different media and tactics. More than ever, marketing communication involves "touchpoints," which are those moments when a possible customer comes into contact with some aspect of your product. A touchpoint can be planned, such as an addition to an Instagram story, or unplanned, such as seeing a friend using the product. Marketing experts Philip Kotler and Kevin Keller recommend several different media and tactics for planned touchpoints: digital and print advertising; sales promotions, such as contests and coupons; special events; PR efforts, such as news releases and speeches; social media; mobile marketing; direct marketing, such as telemarketing and old-fashioned snail mail; and personal selling.

Let's close this segment with a final bit of advice: For all sales and marketing documents, do your research. For class projects and for real-world assignments, make it your business to know everything possible about your product, your competitors and your target audience. Consider using the strategic message planner and/or the creative brief (pages 171–195) for sales and marketing documents. The best, most creative writers are usually those who have done the most research. In sales and marketing—as in all areas of strategic writing—knowledge is power.

# 4B Social Media in Sales and Marketing

## Purpose, Audience and Media

Marketing expert Matt Zilli says that cutting-edge marketing involves a fusion of "fulfilling customer needs, storytelling and digital interactions." Social media can be a powerful ally in each of those marketing areas.

Recent surveys show that the top social media platforms used by advertisers are, in order, Facebook, Instagram, LinkedIn, YouTube, Twitter, Facebook Messenger, Pinterest and Snapchat. However, that's only half the story in social media marketing: Those platforms are the preferences for paid ads and "sponsored" or "promoted" posts. Marketers also use so-called organic messages, which are free updates that organizations post, at no charge, in their own accounts on Facebook, Twitter, Instagram and other social media—think of dictionary company Merriam-Webster's edgy Twitter account, for example. Marketers also can respond to consumers' posts through those accounts.

Marketing analyst Michael Stelzner says that the top five advantages of social media marketing, both paid and organic, are increased exposure of organizations and brands, increased traffic to company websites, generating leads for eventual sales, improved sales and developing loyal followers. Note that most of those reasons don't involve immediate sales. Much of social media marketing focuses on developing relationships that eventually lead to sales.

Economists estimate that receipts from online shopping in the United States will approach $600 billion by 2024. Interacting with customers where they shop—online, in other words—will continue to gain importance for marketers.

## Content of Social Media Marketing Messages

Despite the boom in online shopping, the Content Marketing Institute recommends limiting "buy now" requests to just one-third of your social media marketing messages. The next section in this book (4C: Content Marketing) reviews marketing messages that build relationships that, ideally, lead to sales. In the early 2020s, the hottest content in social media marketing is video: At the beginning of the decade, more than 70% of marketers said they planned to increase their use of video in social media marketing. YouTube remains the top platform for those efforts, followed by Facebook and Instagram. Most videos are in the one- to three-minute range, and most are horizontal, as opposed to vertical or square.

In addition to building relationships with potential customers, another goal of social media marketing is to gather information about consumers. Databases that store such information now drive much of marketing communication—and social media can be a highly effective tool for gathering facts about current customers as well as potential customers. Information-gathering tactics can include giveaways of products, including music downloads, research reports and related items; surveys and questionnaires that also gather demographic, psychographic and behavioral information with the promise of sharing the results or entering the responder into a merchandise drawing; contests that involve consumer-generated content, such as anecdotes or photographs; and discounts in the forms of written or scannable codes. Again, social media can be ideal for reaching consumers with such tactics.

Social media marketers also need to join online, consumer-initiated conversations about their brands. Research shows that almost three-fourths of consumers say that social media reviews and comments affect their purchase decisions and their opinions about products. Direct social media interaction between consumers and organizations also has been shown to improve consumers' impressions of those organizations.

Finally, technology is rapidly changing the way strategic writers create, target, schedule and send social media marketing messages. In some organizations, customer data is stored in stand-alone CRM (customer relationship management) databases. New technology, however, is allowing marketers to merge CRM databases with online social media dashboards that can help target, generate, review, schedule and send individual social media messages based on information in CRM databases. These merged programs are known as DMPs (data management programs) and are offered by marketing-service companies such as Hootsuite, Sprout Social and Agorapulse.

# 4C Content Marketing

## Purpose, Audience and Media

According to the Content Marketing Institute, "Content marketing is a strategic marketing approach focused on creating and distributing valuable, relevant and consistent content to attract and retain a clearly-defined audience—and, ultimately, to drive profitable customer action."

For strategic writers, content marketing means positioning your organization as a supplier of information that customers and potential customers want and need (even if they don't know that yet). This useful information is often entertaining and emotional. It frequently involves storytelling. It always is designed to position your organization as an interesting, trusted, welcome friend. Content marketing begins with an intense focus on target markets and what they want.

Successful content marketing builds relationships with target markets and increases their loyalty to your organization. And in marketing, customer loyalty is huge: Experts estimate that attracting a new customer costs five times as much as keeping a current one. Reducing the number of lost customers by just 5% a year can increase annual profits by 25% to 100%. Besides prospecting for new customers, successful marketing programs often focus on building current-customer loyalty, knowing that profits will follow.

The media for content marketing are those preferred by the target markets. Content marketing often uses social media such as Instagram and Twitter to feature or link to multimedia storytelling efforts or specific landing pages within your organization's website.

## Examples of Content Marketing

Content marketing involves ideas and stories (loyalty-building content) delivered through an ever-expanding variety of media: magazines, both paper and digital; virtual reality presentations (for consumers with headsets); videos; multimedia narratives (online written stories with embedded photos and/or videos); email newsletters; livestreams; slideshows; webinars; digital games; social media posts; specialized websites (often called microsites); podcasts; apps; blogs; texts; hashtag campaigns; e-books; and so much more.

And, in content marketing, what messages fill those media? So-called hacks and how-to instructions; recipes; compilations of consumer-generated content (such as shared advice and/or photos); multimedia stories on topics of interest (including stories about your organization and its employees); entertaining or informative images or infographics; white papers (fact-filled reports); case studies; interviews with interesting people; real-time question-and-answer sessions; news about popular products; concert livestreams; and much more. All are designed to build relationships, increase contacts and enrich brand image.

Specific examples of successful content marketing include:

- "Will It Blend?" Blendtec videos on YouTube
- Trader Joe's online "Fearless Flyer" newsletter (www.traderjoes.com/fearless-flyer)
- Red Bull's *Red Bulletin* digital magazine (www.redbull.com/us-en/theredbulletin)
- Netflix "Queue" podcasts via Stitcher
- Arby's Twitter account (@Arbys)

None of these examples features a strategy of "buy, buy, buy." Rather, the strategy is "inform and entertain."

Several of the sections in this book offer instructions for creating content marketing tactics: microblogs and status updates; blogs; podcasts; video news releases and direct-to-audience videos; newsletter and magazine stories; audio advertisements; video advertisements; mobile messages; brochures; and business reports.

## CONTENT MARKETING TIPS

1. **Begin with research**: Who are your target markets? What do their members want? How do you know that? What can your organization offer them that others can't?
2. **Be strategic**: Content marketing is goal-oriented. Every story your organization chooses to tell should have a specific goal and should complement other stories you've told. Content marketers say their top challenge is coordinating campaigns that involve several brands within an organization.
3. **Measure success**: Research which specific content-marketing tactics are turning followers into customers. Research which messages are generating increased sales to current customers.
4. **Stay current**: A good starting place to review successful ideas for content marketing is the Content Marketing Institute (contentmarketinginstitute.com).

# 4D Social Media Calendars

## Purpose, Audience and Media

A social media calendar schedules the distribution of goal-oriented social media posts. In fact, a social media calendar often is part of a larger online social media dashboard that can automatically post your pre-prepared social media messages. (Computer dashboards are software programs, often web-based, that provide key information and programming opportunities for particular topics and functions.) Social media dashboards also can help you monitor engagement, which measures how and how much your target audience is responding to your posts. Social media calendars can help you create well-organized, sustained, on-message social media campaigns.

The primary audience for your social media calendar is your colleagues, supervisors and clients. A social media calendar can help your team agree on *what* you want to say, *which* media you'll use and *when* you'll send each post. Many social media dashboards allow you to enter proposed social media posts days or weeks before the actual posting. Those dashboards then allow other members of your team to review and approve the posts before the dashboard will release them at the scheduled time. Your target audiences will receive your messages (we hope!), but they'll probably never see your actual social media calendar.

Social media calendars can exist on paper (often for team meetings), but most exist online as part of larger social media dashboards.

### KEY TO SUCCESS

Social media calendars should feature messages that blend your organization's goals with your target audiences' preferences regarding social media platforms and content, such as informative videos or links to entertaining articles.

## Content

Social media calendars are strongly related to integrated marketing communications and to content marketing. As you'll recall, one goal of integrated marketing communications (pages 42–43) is to help an organization stay on message—to use many different communication tactics and channels to deliver one clear message to a specified target audience. That's also the goal of a social media calendar: It helps you use social media platforms such as Facebook, Twitter and Instagram to send consistent, non-contradictory messages to your target audience or audiences. Social media calendars are an important part of planned communication.

Social media calendars also are essential to content marketing (pages 245–246), which seeks to build relationships with customers and potential customers by positioning your organization as a regular, trusted supplier of interesting, entertaining and useful information. Your calendar can help you plan and sustain a steady stream of engaging, audience-oriented posts.

Creating and managing a social media calendar involves a familiar four-step process in strategic writing: research, plan, communicate and evaluate.

1. **Research**: Again, whom do you want to communicate with? Why? And what do they—or might they—want from you? What are their favorite social media platforms? Shannon Tien of Hootsuite, a leading supplier of social media dashboards, recommends that you conduct a social media audit before creating a social media calendar. A social media audit can include assessments of goals, current use of social media (which platforms does your organization now use, and which are most successful?), which kinds of posts are generating the most engagement and who is most qualified to lead the new social media program.
2. **Plan**: Most social media calendars use a grid, such as the example on page 251, to show the specific details of each social media post—everything from timing to message content.
3. **Communicate**: Again, your social media calendar can be part of a social media dashboard that automatically sends your messages. Dashboards can allow you to communicate with team members and gain approval before each message is launched.
4. **Evaluate**: Social media dashboards also can help you discover which specific messages draw the most engagement with your target audience or audiences. For example, did Instagram outperform Twitter? Did "how-to" videos outperform links to relevant articles? Which days and which times of day drew the most engagement? You can use those findings to improve your ongoing social media campaign.

A concept called content curation can help you create effective posts for your social media calendar. Initially, content curation involves searching (usually in the digital world) for stories, images, memes, videos, websites and more that your own target audiences would enjoy. You can direct your audiences to this interesting, useful and perhaps entertaining material through your own posts. After discovering relevant material, content curation involves organizing those sources, adding your own original creations to the collection and having a rich library of relevant material for your ongoing social media campaigns. Finally, content curation involves constantly replenishing and refining that library.

## Organization and Format

If you Google "social media calendar examples," you'll quickly find dozens of approaches to managing the contents and organization of an ongoing social media campaign. You'll also discover that most social media calendars specify three things:

1. **Where?** In other words, which platforms will you use? Facebook? Twitter? Instagram? Which combination of platforms will be most effective in reaching your desired target audiences?
2. **When?** What specific day and what specific time of that day will each post launch? Organizations such as Sprout Social offer free data on the best times to post within each social media platform. For example, in 2020, the best time for a nonprofit organization to post on Instagram was Wednesday at 2 p.m. Those posts drew the most viewer engagement.
3. **What?** What will you post? What words? What images, including videos? What hashtags? What links? Obviously, this section of a social media calendar specifies content. Some calendars include the full post; others include just a concise description of each post. Many calendars specify the category of each post, such as "User Generated Content" or "How To" or "Survey." Some organizations give each weekday a theme—for example, Mondays are for recipes, Tuesdays are for frequently asked questions and so on.

A good social media calendar helps ensure a good, on-message social media campaign. And a good social media campaign can help ensure successful, productive relationships with important target audiences.

## Social Media Calendar Tips

1. **Be frequent**: How many times a day should you post on each platform? Expert opinion varies, so do your research. General advice suggests one to two daily posts for Facebook; one to three for Instagram; and anywhere from three to 50 for Twitter (most sources recommend three to 10 daily tweets). As your organization gains experience and studies engagement with your audiences, you can refine the frequency of your posts.

2. **Don't kill spontaneity**: Social media calendars shouldn't restrict your ability to respond quickly and effectively to unforeseen problems and opportunities. For example, @ Oreo (Oreo Cookies) will forever be in the Twitter Hall of Fame for its memorable tweet when the power failed in the stadium at Super Bowl XLVII in New Orleans: "Power out? No problem. You can still dunk in the dark." No social media calendar can anticipate a moment like that.
3. **React to changes**: Sticking with your social media calendar in times of sudden crisis for your organization or for its key audiences might make your organization seem out-of-touch or, worse, uncaring. A good crisis communications plan should include directions for rapidly revising a social media campaign. That might involve editing or canceling previously scheduled posts. A national organization once had to apologize to its members about a post promoting a convention that just hours before had been canceled because of a hurricane.
4. **Celebrate**!: In the United States in late November, you might consider a Thanksgiving theme for some of your posts—ditto for early February and Groundhog Day. Be mindful of traditional holidays, and consider seeking out some unusual ones. Did you know that Jan. 13 is "Clean Off Your Desk Day"? And March 14 is, of course, Pie Day/Pi Day (3.14).
5. **Don't oversell**: Remember that you're using social media to build solid relationships. Ideally, sales or donations will follow. As we note in 1E: Writing for Social Media, the Content Marketing Institute recommends limiting sales pitches to 20% of your posts at most.
6. **Keep current**: Social media platforms and functions evolve, flourish and sometimes die. Remember Vine videos? Keep familiar with what platforms your target audiences prefer.
7. **Build safeguards**: Don't let the same person create, review and distribute your social media posts. Social media dashboards often allow you to implement an approval process to ensure careful review and approval by more than one person before posting.

## Social Media Calendar

| WHERE | WHEN | | WHAT | | IMAGE/VIDEO/LINK |
|---|---|---|---|---|---|
| | Date | Time | Category | Content | |
| **MONDAY: CLINICS** | | | | | |
| **Facebook** | Monday, 9/1 | 10 a.m. | Free Clinic Announcement | Invitation to Tuesday's free blood-pressure clinic. Link to blood-pressure stats. | Photo of nurse with visitor rcmc:bp719.net |
| **Twitter** | Monday, 9/1 | 9 a.m. | Free Clinic Announcement | Invitation to Tuesday's free blood-pressure clinic. | Photo of nurse with visitor |
| | | 3 p.m. | Free Clinic Retweet w/Comment | Response to tweet praising previous blood-pressure clinic. | None |
| | | 7 p.m. | Free Clinic Announcement | Invitation to Tuesday's free blood-pressure clinic. Link to blood pressure stats. | Photo of blood pressure cuff rcmc:bp719.net |
| **Instagram** | Monday, 9/1 | 11 a.m. | Free Clinic Announcement | Invitation to Tuesday's free blood-pressure clinic. | Photo of nurse with visitor |
| | | 7 p.m. | Free Clinic Announcement | Invitation to Tuesday's free blood-pressure clinic. Graphic of five blood-pressure reduction tips. | Graphic of five tips for reducing b-p |
| **TUESDAY: NUTRITION** | | | | | |
| **Facebook** | Tuesday, 9/2 | 10 a.m. | Nutrition Announcement | Invitation to Friday's Olympic skiers nutrition clinic. Link to nutrition stats. | Highlight video of skiers rcmc:nut618.net |
| **Twitter** | Tuesday, 9/2 | 9 a.m. | Nutrition Announcement | Invitation to Friday's Olympic skiers nutrition clinic. | Photo of Olympic skiers |
| | | 3 p.m. | Nutrition Retweet w/Comment | Response to previous tweet praising RCMC nutrition clinics | None |
| | | 7 p.m. | Nutrition Announcement | Invitation to Tuesday's free blood-pressure clinic. Link to nutrition stats. | Highlight video of skiers rcmc:nut618.net |
| **Instagram** | Tuesday, 9/2 | 11 a.m. | Nutrition Announcement | Invitation to Friday's Olympic skiers nutrition clinic. | Photo of Olympic skiers |
| | | 7 p.m. | Nutrition Announcement | Invitation to Friday's Olympic skiers nutrition clinic. Image series: Vegan tacos recipe. | Image series |
| **WEDNESDAY: WEDNESDAY WONDERINGS – Q&A** | | | | | |
| **Facebook** | Wednesday, 9/3 | 10 a.m. | Wednesday Wonderings: Q&A | Answer to Emergency Room question. Solicitation of new questions. | ER photo |
| **Twitter** | Wednesday, 9/3 | 9 a.m. | Wednesday Wonderings: Q&A | Answer to insurance question. Link to RCMC Insurance FAQ page | rcmc:infaq.net |

# 4E Proposals and Marketing Communications Plans

## Purpose, Audience and Media

A proposal is a report-like document that promotes and describes a plan—for a new website, a new departmental structure, a public relations campaign, an elaborate special event and so on. A marketing communications plan is a similar document that clearly outlines an overall integrated promotional strategy and how it will be implemented. Both describe the need for or the advantages of action. After an organization approves a marketing communications proposal, it becomes a plan—a roadmap—that guides all integrated marketing strategies for a specified time period. (See pages 42–43 for a quick review of integrated marketing communications.) In other words, a proposal can evolve into a marketing communications plan. To streamline this book, we'll call this a marketing communications plan, but exactly the same format could be used for a purely public relations proposal or a communications plan not oriented toward consumers and sales.

The audience for a proposal or a marketing communications plan is the person or people with the power to approve and implement the plan. The audience will often be a client or the leaders of an organization.

The primary medium for proposals and marketing communications plans continues to be paper. These documents must be studied; therefore, they generally appear as bound documents. However, formal oral presentations of proposals and plans include multimedia elements such as slide presentations (using PowerPoint, Prezi or similar programs), videos and other visual aids.

### KEY TO SUCCESS

Through clear organization, a successful proposal or marketing communications plan shows how a specific course of action will solve a well-defined problem or will seize a well-defined opportunity.

## Format/Design

Marketing communications plans follow the basic format and organizational structure of proposals. Write proposals on standard-sized (8.5-by-11-inch) paper. Single-space the text. Double-space between paragraphs, and don't indent paragraphs. Number the pages, starting with the executive summary (see page 330); do not number the page(s) of the table of contents. The formats of proposals often include the following design and graphics elements:

- Colorful charts, graphs and other visuals to reinforce and clarify meaning. Proposals can be long and daunting; graphics can highlight and clarify key points and provide visual relief.
- Bold, large type for the title on the title page (18-point Times is a standard size for report titles)
- Section titles and subheadlines in boldfaced type
- Margins of at least one inch
- White space (extra spacing) between sections; in long reports, each new sections often begins on a new page

As you develop your plan, be sure to review Strategic Design (pages 35–41).

## Content and Organization

A formal proposal can contain more than a dozen sections, each of which generally begins on a new page. (Less formal proposals can discard some of the sections.)

Tailor your proposal to the specific situation, but in general your proposal should contain the following sections. The order of presentation below reflects the order of a traditional proposal.

### *Memo or Letter of Transmittal*

With a paper clip, attach a brief memo (see pages 322–327) or letter addressed to your audience to the cover or title page of your proposal. The memo or letter essentially says, "This proposal presents an idea to address the challenge of . . ." [or "to seize the opportunity of . . ."]. It can close with an implicit or explicit call to action, such as "I'm available to discuss this at your convenience." A "Thank you for your time and consideration" can follow the call to action. A memo or letter of transmittal generally is omitted from a marketing communications plan, which often has evolved from an earlier proposal.

Use a memo for proposals submitted to internal audiences (groups within your organization). Use a business letter for proposals submitted to external audiences (groups outside your organization). If you have distributed PDFs of your proposal via email, a concise email message can serve as your memo of transmittal.

## Title Page

The title page often is the cover of the proposal. It includes a title; a descriptive subtitle, if necessary; the name(s) of the author(s); and the date. Your title should be compelling, positive and descriptive. A snappy, teasing title—much like a slogan—is often effective. If your proposal introduces a promotional campaign that has a theme, consider using that theme as your title. A descriptive subtitle can make it clear that the document is a proposal. The subtitle generally includes the word *proposal*. For example:

**PARTNERS FOR PROGRESS**

**A Proposal to Gain Local Consent for the Expansion of the Portland Headquarters**

## Table of Contents

A table of contents lists each section, in order, and the starting page for each of those sections. (Do not list the span of pages for each section; just list the starting page.) In the table of contents, do not include the memo/letter of transmittal, the title page or the table of contents itself. With a headline, clearly label this page as the table of contents.

## Table of Charts and Graphs

This optional section lists the names and page numbers of the charts and graphs in the proposal. This section generally is not listed in the table of contents.

## Executive Summary

An executive summary is a concise, ideally one-page overview of the proposal's highlights. In general, include an executive summary if the report is formal and will take more than 15 minutes to read. In proposals, an executive summary normally summarizes the following sections: situation analysis, target audiences and tactics. Each of these sections is described below.

Do not use your executive summary as an introduction to your proposal. A reader may skip the executive summary. Everything in your executive summary must appear elsewhere in the proposal. You can avoid treating the executive summary as an introduction by writing this section last.

Begin numbering pages with the executive summary.

## *Situation Analysis (the Problem or Opportunity)*

The situation analysis requires in-depth research and describes, in detail, the status quo—the way things are right now. However, the situation analysis presents the status quo in such a way that it fills readers with the desire to act to solve a problem or seize an opportunity. In the situation analysis, *do not* mention the solution (the plan) that the proposal will present. Keep the focus solely on the problem or opportunity. Let your readers, as they study the facts, realize that the situation demands a response.

Ideally, the situation analysis relates to the fulfillment of an important organizational goal or goals and can mention them. In a marketing communications plan, this section also can discuss competitors and their status in the marketplace. After all, they're part of the existing situation.

## *Statement of Purpose*

This brief section announces the purpose of the proposal. Having just finished the situation analysis, your readers ideally are saying, "We need to act. We need to address this situation." The statement of purpose reassures readers by simply stating, "This proposal presents a plan to win citywide acceptance for the expansion of the Portland headquarters."

The statement of purpose often can be one clear, confident sentence.

## *Target Audiences*

Much of your work as a strategic writer involves creating productive relationships with specific audiences. Proposals for public relations, advertising, sales and marketing and business communications generally involve plans designed to affect relationships. Thus, proposals usually specify target audiences.

A section on target audiences generally begins with a brief explanatory paragraph. This first paragraph explains that the forthcoming plan focuses on a clear target audience or a set of target audiences. This first paragraph can be as short as one sentence: "The forthcoming plan targets five distinct audiences: the City Commission, neighborhood committees, city religious organizations, city business leaders and the local news media."

After presenting this brief explanation, describe each target audience. Descriptions should include demographic information (non-attitudinal information such as age range, gender, income, race and education), psychographic information (attitudinal information such as political philosophy, religious beliefs and other important values) and behavioral information (information about traditional actions and habits). Descriptions also should include the desired resources that each public controls and the nature of their relationship, if any, with your organization.

Use internal headlines to provide a new subsection for each new target audience.

## *The Plan (Goals, Objectives, Strategies and Tactics)*

Plans usually exist in an outline form and begin with a goal. A goal is a general statement of the outcome you hope your plan will achieve. Goals often begin with infinitives, such as "To improve" or "To increase." By beginning your plan with a verb, you place an immediate focus on action. For example, a goal might be "To win citywide acceptance for the expansion of the Portland headquarters." The goal often echoes the earlier statement of purpose: "To win citywide acceptance for the expansion of the Portland headquarters." Some plans have more than one goal. In such plans, each goal would have its own set of objectives.

After the goals come the objectives. Unlike goals, objectives are specific. Objectives clarify the exact things you must achieve to reach a goal.

How many objectives should your plan have? In a plan with more than one target audience, consider presenting one objective for every target audience. Each objective would specify what outcome you hope to achieve with each audience. For example, an objective related to the earlier goal might be "To gain approval from the City Commission by June 15."

Like goals, objectives usually begin with an infinitive. Objectives also are measurable; that is, they establish a clear line between success and failure. Number your objectives and include a deadline.

Some plans include Strategies between the Objectives and Tactics. A strategy is a general description of the tactics you'll implement to fulfill an objective. In other words, strategies help you move from specific objectives to specific tactics. For example, one strategy under our City Council objective might be "Contact individual council members with information about the expansion." Beneath, list specific tactics that provide precise details about how you propose to achieve that particular tactic.

Tactics are the actions you recommend to achieve each objective. List and describe tactics under the appropriate objective. Unlike goals and objectives, tactics don't begin with infinitives. Tactics begin with active verbs; they're commands. For each tactic, include the following information: brief description, deadline, budget, special requirements, supervisor and evaluation. Number the tactics under each objective, beginning with "Tactic #1" for each new objective. Because we generally want to reach each targeted public through messages in multiple channels that reinforce one another, it's likely that each objective will have more than one tactic.

The beginning of a plan, therefore, looks something like this:

**Goal**: To win citywide acceptance for the expansion of the Portland headquarters

**Objective #1**: To gain approval from the City Commission by June 15

**Strategy #1**: Contact individual council members with information about the expansion.

**Tactic #1**: Send personal letter from CEO Sarah Jones to each city commissioner

*Brief description:* Ms. Jones will send each commissioner a personal letter announcing the proposed expansion of the Portland headquarters. The letter will emphasize the benefits to the city.

*Deadline:* March 31

*Budget:* $12 for stationery and postage

*Special requirements:* City Commissioner Dennis Jackson is blind. He prefers to receive correspondence via email. His computer has audio-reader software.

*Supervisor:* Communications Specialist Kris Palmer

*Evaluation:* The success of this tactic will be measured by whether the City Commission is already familiar with our key points when we make our formal presentation.

Ideally, the plan leaves readers with no questions about the details. Specify sizes, colors, dates, places, prices and so on. If the description of tactics becomes too detailed and begins to clutter the plan, consider including samples or sketches in a Supplements section. If you include supplements, direct the readers' attention to that section at appropriate places in your description of each tactic.

## Timetable

This section is optional because the plan's objectives and tactics already include proposed deadlines. As a separate section, however, a timetable can be a useful chart. Organize the timetable in chronological order, with the first action first and the last action last. Place the date in a left-hand column and the related action in a right-hand column:

| | |
|---|---|
| March 31: | Mail CEO letters to city commissioners. |
| April 1: | Post first podcast. |
| April 4: | Send open-house invitations to neighborhood committees. |

## Budget

Although you already have specified the cost of each tactic, include a "line-item" budget in your proposal. In a two-column format, list each expense and the projected cost. In a detailed, lengthy budget, consider including a "contingency line" for unforeseen expenses. To include a contingency line, total all the expenses and label the resulting sum as a subtotal. Next, determine what 10% of the subtotal would be and list that amount in your contingency line. (Check with your supervisor or consult previous proposal budgets to determine an acceptable percentage of the subtotal for your contingency line.) The end of a detailed line-item budget would look like this:

| | |
|---|---|
| Posters, 40 copies | $200.00 |
| Campaign buttons | 150.00 |
| New website section | 825.00 |
| SUBTOTAL | 1,175.00 |
| Contingency budget | 117.50 |
| TOTAL | $1,292.50 |

## Challenges

This optional (and generally rare) section presents and refutes challenges to the situation analysis and the proposed tactics. Consider including it when your proposal contains controversial material, and, therefore, obvious challenges exist. Clearly and concisely state the challenge and the rebuttal:

Challenge: The City Commission opposed the past four corporate building proposals.

Refutation: Those four proposals did not include an expanded workforce. Our proposal includes more than 200 new jobs for the Portland area.

## Additional Benefits

This optional section details the "add-on" benefits that your proposal would create—besides the basic benefit(s) of reaching the identified goal(s). For example, improving a relationship with a target public might have additional, future benefits. List those benefits as "Additional Benefit #1" and so on.

## *Conclusion*

Proposal conclusions are brief. Summarize the need for action. State that the proposal offers a plan to address that need. Consider closing with specific recommendations for next steps, which might include dates and places for future discussions of the proposal and procedures for the formal acceptance of the proposal. Consider recommending a timetable for those actions. Some proposals eliminate this section, preferring to close with the plan itself. The conclusion of a marketing communications plan might concisely remind your communications team of what success will look like.

## *Supplements*

This optional section can include samples, dummy layouts, charts, graphs and articles—anything that the proposal calls for or that supports the clarity or persuasiveness of the proposal. Include only materials cited earlier in the proposal.

### PROPOSAL/PLAN TIPS

1. **Be diplomatic**: Don't present your proposal or plan as the savior of a sinking ship or as the solution to stupid errors. Your proposal often will target an area managed by the very people who will evaluate the proposal. Don't hurt their feelings or make them defensive. Such diplomacy is particularly important in the situation analysis.
2. **Bring target audiences to life**: Proposals for advertising campaigns sometimes use first-person narratives instead of formal descriptions of target audiences. For example, in an ad-campaign proposal, a description of a target audience might begin "My name is Michael Khomsi, and I'm a 27-year-old Arab American. I grew up in Detroit. . . ." Such a description, though fictional, would be a highly detailed analysis of a representative target consumer.
3. **Consider a unifying theme**: If you propose a campaign that has a theme, incorporate it throughout the document where appropriate, especially in the Statement of Purpose section. You might even label that section Statement of Purpose and Theme. The theme also can be the title of the proposal.
4. **Focus on outcomes**: In tactics, be sure to evaluate outcome rather than process or output. For example, the evaluation measure for news releases should not address the number of news releases you distribute. Rather, the evaluation measure should address how many media outlets published or broadcast the main points of the news release—or, better, whether the media's audiences received and believed your message.

## PROPOSAL/PLAN TIPS

5. **Be concise**: Avoid the temptation to pad a proposal or marketing communications plan with wordiness, useless information or unneeded sections. Concise documents show respect for their audience. These documents should be long enough to thoroughly fulfill their purpose—and no longer.
6. **Seek advice**: If possible, show drafts of your proposal or marketing communications plan to your manager as you progress. These are important documents. Avoid distributing an unedited, unapproved proposal or plan.
7. **Communicate with team members**: Proposals and marketing communications plans often are written by teams. If you are the team leader, assign specific tasks to individuals. Assign a deadline for each task. Make these assignments in writing—in a memo or email—to avoid any misunderstandings. If the document will take several days or weeks to complete, hold quick progress meetings or ask that members of your team send you periodic progress reports.
8. **Make it attractive**: Appearances count. Consider working with an art director to create an attractive (though economical) proposal.

# 4F Mobile Messages

## Purpose, Audience and Media

Mobile messaging is a key component of mobile marketing. In the words of the Mobile Marketing Association, mobile marketing is the promotion of goods and services by means of "advertising, apps, messaging, mCommerce and CRM (customer relationship management) on all mobile devices including phones and tablets." But don't think only of retailers when you think of mobile marketing messages: Schools, sports teams, civic organizations and potential employers effectively use this form of communication to deliver messages. Here are some examples of how mobile marketing can work:

- Text messages
- Text coupons
- Downloadable software apps that offer special services
- Microblog messages, such as tweets, about product reviews, availability or discounts
- GPS (global positioning systems) technology that issues coupons or other sales messages when customers are near their favorite stores or restaurants

This segment focuses on three similar aspects of mobile messaging:

- SMS (or short message service) messages, which are limited to words
- RCS (or rich communication services) messages, which are multimedia and interactive. Experts predict the replacement of most SMS messages by RCS messages by 2025.
- In-app messages (text messages that usually promote a particular app feature), which generally are limited to words

In the best traditions of strategic communication, these marketing messages should focus on individual wants and needs. Under the U.S. Telephone Consumer Protection Act (TCPA), the recipient must have requested such messages by responding to an initial text or email or through the organization's website: TCPA requires written consent from a recipient before a messaging campaign can begin. The same act specifies that you must offer recipients an easy way to opt out of such messages. Apps often contain Preference sections in which users can specify what kind of in-app messages, if any, they wish to receive. A database can help ensure that your messages directly target an individual's known interests. Research shows that teenagers and young adults are the most receptive audiences for sales-related text messages, but the practice is rapidly spreading to older publics.

College students already know that texting is big business, but it's still impressive to learn that, even back in 2020, more than 90% of U.S. consumers said they had received a business-related text. That was a jump of 20% from 2019.

Mobile messages are increasingly important in strategic communication: Tablets and phones outsell laptop computers; usage of popular messaging apps now exceeds usage of popular social media networks; and texts are four times more likely to be opened than emails. A recent study by Zipwhip, a supplier of texting-management systems, found that most consumers have almost 200 unread emails in their inboxes—but most have either one or zero unread texts in their phones.

**KEY TO SUCCESS**

Mobile marketing messages are highly targeted, concise, have a strong visual appeal (in functions that allow visuals) and often lead receivers to a web-based landing page. However, consumers say the most valuable business-oriented texts are reminders of appointments.

## Format/Design

Business-related texts are brief: SMS texts are limited to 160 characters and spaces. RCS texts can include visual elements and interactivity, and they do not have a restriction regarding character count (though brevity is important). In-app messages generally begin with the app's identifying icon and then a brief text message. With user permission, in-app messages can appear when a phone or tablet is reactivated, even if the app is not open.

Message formats and designs—as well as payment plans—are affected by the texting service provider and technology that an organization selects.

## Content and Organization

Successful mobile messaging programs encourage interaction and continuing conversations: For example, a successful promotion that leads to a purchase could be followed by a thank-you message; and then a request to rate the product; and then a coupon for a related purchase. Companies use databases, which can link to texting-management systems, to keep track of interactions with specific customers (see Integrated Marketing

Communications, pages 42–43). In addition to such basics as name, phone number and kind of mobile device, such databases often include a customer's "RFM" data, meaning *recency* (when did the customer last purchase the product?), *frequency* (how often does the customer purchase the product?) and *money* (how much does the customer spend for each purchase?).

Because mobile marketing messages are brief, you should focus on known recipient interests and attractive keywords, such as *free, save* and *discount* (see page 200). Despite their variety, successful sales-related text messages share these qualities:

- They are brief.
- They focus on benefits.
- They promote a continuation of the relationship with the consumer; they strive to prompt a response.
- They include a call to action on the recipient's part.
- They include a link, generally to a sales-oriented landing page.

Two categories of sales-related texts exist: initiated messages and response messages. Initiated messages often strive to attract recipients to points of purchase such as websites or traditional brick-and-mortar businesses. Such messages might read: "40% savings and free shipping at mystore721.com! Exp. 3/6/23" or "Show this text for 10% discount on next purchase at MyStore in Center City! Code: 5XAB9." (Entered into your database, the code would prevent the consumer from using the discount more than once.) Sales-related texts include thank-you notes, shipping updates, requests for product reviews and even billing.

Response messages are replies to messages initiated by consumers. For example, a consumer might use a phone app to make a Jan. 12 reservation at the XYZ Café. A sales-related response text might read "Thank you! Jan. 12 reservation confirmed for 1 p.m. Show this text for 10% discount on any XYZ meal purchased in February. Code: 9GKL1."

Identify your organization in mobile messages, which often can begin simply with your organization's name in all caps: "QRSBOOKS.COM: We're shipping your order 24 hrs early! Expect arrival this Tuesday. Reply for 10% off your next purchase." (We all know the frustration of receiving anonymous texts.)

Your mobile messages should strive to create consumer responses. For example, a bookstore might send this SMS message: "Text BOOKS to 55555 to see the latest discounts on your favorite authors at QRSBooks.com!" Note that the designated keyword (BOOKS) is easy to see and easy to text.

A similar RCS message could include a photo of a popular book cover, perhaps a new work by the recipient's favorite author. And the RCS message could include an attractive coupon or "Order Here" button instead of the less popular "short code" option. New studies show that consumers are tiring of texting so-called short codes such as BOOKS. Consumers now prefer easy interactivity or quick text exchanges with live human beings (as opposed to business-owned chatbots) over code use.

Studies show that because RCS messages include a strong visual element and easier interactivity, they are more likely to prompt a recipient's response than are SMS messages.

Consult your own phones. What mobile marketing messages do you receive? What makes them effective—or the opposite?

Although most consumers are familiar with their texting payment plans, the Mobile Marketing Association recommends adding this closing to any sales-related texts that solicit a texted response: "Msg&data rates may apply."

## MOBILE MESSAGE TIPS

1. **Integrate**: Be sure to coordinate mobile marketing campaigns with other sales and marketing efforts (see Integrated Marketing Communications, pages 42–43). Mobile marketing campaigns should reflect the strategies and values of an organization's overall marketing program.
2. **Don't oversell**: The Content Marketing Institute recommends restricting direct sales messages to just one-third of your total mobile messages. The remainder should be entertaining and informative. Research from Marketo Marketing shows that excessive, irrelevant messages are the top reason why users cancel in-app communication functions.
3. **Be accessible**: Ensure that your organization's phone numbers are text-enabled. Zipwhip reports that almost one-third of consumers say they have texted a company and not received a reply.

# 4G Sales Messages and E-Blasts

## Purpose, Audience and Media

A sales message is a business message, such as a text, email or letter, that attempts to persuade the recipient to buy a product (a good or a service). Sales letters can be expanded into more elaborate direct-mail packages. An e-blast is an email message that often has a similar goal, but e-blasts can have a variety of other purposes, such as delivering newsletters or product information to recipients. If a professor uses email to inform you and your classmates about a new assignment, that's an e-blast.

Sales letters and e-blasts are mass-produced. However, that mass production shouldn't prevent you from including knowledge about the individual recipient. Increasingly, companies have detailed customer-information databases that allow them to send highly personalized sales messages through a variety of media. Organizations also can purchase detailed lists of potential customers or donors from list brokers. Ideally, however, recipients of e-blasts "opt in." In other words, they previously agree to receive your email messages; they often grant such permission when making an online purchase. In the United States, unsolicited sales e-blasts may violate the law unless they contain a clear "unsubscribe" option. In the European Union, failure to secure previous permission for e-blasts violates the law.

The audience of a sales letter or e-blast is one person. Again, even though you're probably sending dozens, hundreds or thousands of similar messages, attempt to personalize each one. Never, for example, send a sales letter to "Dear Resident." (See 3C: Strategic Message Planners for the kinds of information to gather about consumers and other important target audiences.)

E-blasts, of course, use email delivery. Sales letters—often as expanded direct-mail packages (pages 268–269)—still use paper and are delivered through the postal service. Do paper letters still work in our digital world? If such letters are well-targeted, research shows that many recipients find them more personal than sales approaches via phones, tablets and laptops. The old-school approach of paper letters actually can gain a recipient's attention in our world of online clutter.

**KEY TO SUCCESS**

Successful sales messages and e-blasts rely on database-driven knowledge of individual consumer preferences.

## Format/Design

For sales letters, follow the general guidelines for the business-letter format on pages 297–301. However, sales letters often delete the recipient information from the heading—the three lines that include the recipient's name, address and city, state and ZIP code. They do include the date.

Unlike most other business letters, sales letters highlight key passages with design elements such as boldface type, different-colored type, underlining, capital letters, subheadlines—and even handwritten sticky notes, prepared and attached to the letter by a machine. The signature at the bottom of a sales letter often is overprinted in blue ink to make the letter seem hand signed. The P.S. also may be handwritten in blue ink.

E-blasts generally include colorful images (such as coupons), links to websites or even promotional videos to encourage recipients to learn more about the product. E-blast newsletters contain "unsubscribe" and social media sharing options, usually at the bottom.

## Content and Organization

Sales-related e-blasts from companies such as Amazon or TravelSmith feature colorful coupons that reflect the recipient's history of purchases. Coupon headlines generally highlight a benefit ("Need to Cure the Winter Blahs?"), a deadline ("Two Days Only!") or a price reduction ("20% Off Your Favorite Designer Labels!"). Such coupons also serve as links to the sales landing page. Other e-blasts take the form of a newsletter table of contents, with each story description including a compelling image, a headline, a punchy one-sentence summary and a "Read More >>" link that takes recipients to the full story.

Like coupon headlines, subject lines for sales-related e-blasts focus on benefits ("Big savings on last-minute gifts!"), deadlines ("Only 48 hours left!") or price reductions ("40% savings and free shipping!"). Subject lines for e-blast newsletters often focus on time-span and the lead story: "Weekly update: Stock prices plunge." The subject line must be powerful enough to capture the recipient's attention and entice them into opening the email.

Sales letters can use a variety of organizational strategies. Such letters often exceed one page. (The theory is that if the letter delivers enough product-related benefits, the recipient will keep reading.) Traditional sales letters feature a six-part organizational strategy.

## *Part 1 of 6: Begin With a Teaser Headline (Optional)*

In sales letters, teaser headlines are optional. A teaser headline generally appears in the upper-left corner of the page, above the date and the salutation. (Remember that sales letters often delete the three lines in the heading that specify the recipient's name, address, city, state and ZIP code). Teaser headlines usually use a different typeface—and often a different color—from the rest of the type. They are larger (usually 18- or 24-point type), and they don't extend across the entire page. Instead, they split into two or three lines and remain in the upper-left corner.

Unlike newspaper headlines, teaser headlines usually don't tell; instead, they tease. They ask a question, mention a problem, state an eye-popping statistic, refer to a solution or highlight words that sell. (For a list of words that sell, such as *free*, see page 202.) The goal of a teaser headline is to capture the reader's attention and get them to read the letter to learn about the headline.

## *Part 2 of 6: Create a Sense of Need or Desire*

In the first paragraph, don't start by mentioning your product or by asking for the sale. Instead, create a sense of need or desire. This often means creating a concise scenario that presents a familiar problem or desire to the recipient. The goal of this section is to remind the recipient that something in their life needs to be better. Don't mention your product yet. The recipient will view the product as a solution—and at this point, you want them to think only about their problem. Keep them focused only on their sense of need. By the time they read the final sentence of this section, the recipient should be filled with a desire to improve some aspect of their life. Note that the sales letter (like the bad-news letter) does not follow the tradition of using the first paragraph to tell the recipient why they're reading the letter.

Particularly in letters that accompany direct-mail packages, this section can be more than one paragraph. However, don't dwell too long on the problem; once the recipient realizes that the situation applies to them, they'll be seeking the solution you offer.

## *Part 3 of 6: Present Your Product as the Solution*

Beginning with a new paragraph, satisfy the recipient by presenting a solution to their problem: your product. Be specific about how your product solves the problem and improves the recipient's life. Discuss the benefits of your product in detail. Remember that a benefit is a product characteristic—a feature—that creates something advantageous and desirable in the recipient's life.

The discussion of your product's benefits often exceeds one paragraph. As noted in the previous Format/Design segment, this portion often features design elements such as boldface type, different-colored type, underlining and capital letters. Consider using such elements to highlight particular benefits. In the longer sales letters that accompany direct-mail packages, this benefit section can continue for several paragraphs.

## Part 4 of 6: Ask for the Sale

In a new paragraph, ask for the sale. Or demand it: "Order yours today! It's easy. Just. . . ." This concise passage is known as a "call to action." Be sure to give all the details about how the recipient can acquire your product. In a short sales letter, this section usually is one paragraph.

If you're concerned that your product's price may dampen the recipient's enthusiasm, consider these ideas from direct-mail expert George Duncan:

- Offer a guarantee or a free-trial period.
- Compare the product's price to the price of something familiar and desirable, such as dinner with friends or a cup of coffee every day for a month.
- Create a sense of urgency with a special benefit: "And if you respond within the next 30 days, we'll also send you a deluxe. . . ."

## Part 5 of 6: Re-evoke the Sense of Need or Desire and Again Ask for the Sale

In case the recipient has become too relaxed, in a new, final paragraph return to the idea you developed in the first paragraph: Something is missing in their life. After briefly re-evoking that sense of need, tell them to purchase your product today. Include a standard "Sincerely" sign-off.

## Part 6 of 6: Add a P.S.

Almost all sales letters add a postscript, a *P.S.*, below the sender's signature and title. (Include an extra space, just as if the P.S. were a new paragraph.) Some postscripts even appear to be handwritten, as if they were an urgent personal note from the sender to the recipient. The P.S. presents one final incentive to purchase the product. A sales-letter P.S. usually describes an additional benefit of the product or, more often, presents a bonus for purchasing soon: "Call now, and you'll also receive a. . . ."

A P.S. is not signed. It appears at the bottom of a letter and is introduced simply by the initials P.S.

## Direct-Mail Packages

A direct-mail package is an unsolicited persuasive message sent to consumers on a mailing list. It attempts to change attitudes, beliefs or actions. Direct mail can be used to raise money, expand membership, educate recipients, generate income, increase renewals or sell subscriptions or products.

Like sales letters, the usual medium for direct-mail packages remains traditional paper and the postal service. Increasingly, marketers use email messages and newsletters to stay in touch with prospects on a regular basis. Most direct-mail packages try to capture an email address and then follow up with email messages.

Each part of a traditional direct-mail package has a specific function. The more pieces in the package, the greater the likelihood that the recipient will look at one of them. The pieces can include:

- An outer envelope—attracts attention with teaser copy and gets the package opened
- A sales letter—explains the offer in detail, sells the benefits to the recipient and asks for the sale
- A brochure—restates the offer made in the sales letter in a visual presentation
- Other teaser devices such as a lift letter (testimonial), product sample, free gift or coupon—attract attention and encourage the recipient to spend time with the package
- A reply card—asks for the sale and tells the recipient how to respond

## SALES MESSAGE TIPS

1. **Understand the recipient**: Focus on the recipient's self-interest. They don't care what you think about the product. They want to know what it can do for them. Any product features that you describe should be presented as benefits to them.
2. **Don't overdo it**: Avoid excessively negative scenarios in Parts 2 and 5. Don't threaten the recipient. If you successfully describe an unpleasant scenario, their imagination will supply the unfortunate consequences of inaction.
3. **Coordinate length and benefits**: Some sales letters, especially those in direct-mail packages that contain brochures and other items, are longer than one page. The theory is that if the recipient clearly sees how they benefit from your product, they'll continue to read for that length.
4. **Use an SMP or creative brief**: Like advertisements, sales letters should present a clear, concise, beneficial image of your product. Consider completing a strategic message planner or creative brief (pages 171–195) before you write a sales letter.

## E-BLAST TIPS

1. **Gain recipients' attention**: A compelling subject line is essential. Ask a question, pique recipients' curiosity or offer a challenge.
2. **Experiment**: Test different subject lines to see which one pulls best. Always measure your e-blast results.
3. **Personalize**: Create e-blasts that are personal and don't appear to be spam. Frequently, content should be unique and informative rather than promotional.
4. **Keep it legal**: Ensure that every recipient on your list has given permission to receive emails from your organization. Always include an "opt-in" option to ensure that your customers want to continue to receive emails. Have an obvious link where customers can unsubscribe.
5. **Build identity**: Keep your organization's style consistent in all e-blasts. The e-blasts should all feature the same colors, logo and format to reinforce your organization's image.
6. **Promote sharing**: Include social media sharing links within your e-blasts so your recipients can easily send them to others, thus increasing your reach.
7. **Study e-blasts that you receive**: As you know, e-blasts are not limited to sales messages. The term extends to any mass emailing to an "opt-in" list of recipients. Nonprofit organizations use e-blasts to inform members and donors of upcoming events. Professional associations often use e-blasts to distribute newsletters and breaking news. Despite such diversity, successful e-blasts have this in common: They quickly and clearly target the interests of the recipients.

## Sales Letter

2010 Ridglea Dr. • Austin, TX 55111 • 555-999-5555

*Hey, gamer!*
*So you think you've seen it all?*

October 26, 2022

Dear Mr. Trip:

Great graphics? Hey, they're everywhere now. Wall-shaking audio? What game doesn't have that? Hidden features and secret levels? Big deal—nothing new there. Aren't there any surprises left for serious gamers?

*Starklight Random* is the new game from David Smith. You know David as the award-winning creator of *Night Terror* and *Are You Sleeping?* When you load *Starklight Random* the first time, the graphics and audio will dazzle you. But when you launch it the second time, they'll stun you: They're different. Our unique randomizer technology creates a new game every time you play. New graphics. New audio. New challenges. Goodbye to the same-old, same-old: *Starklight Random* contains **MORE THAN ONE THOUSAND VARIATIONS.** No more waiting for the latest version of your favorite game. (And, *yes*, you can save each version—game by game.)

Mr. Trip, we want you to be one of the first players of *Starklight Random.* In just one week, you could be playing the 1,000+ variations for the special introductory price of only $79.95. That's $20 off the store-shelf price—**half the cost of a pair of high-end headphones.** Log on to MGSintgames.com. We accept all major credit cards, and we'll ship *Starklight Random* the day we receive your order.

OK, it's time to head back to the game console. Go play the old, familiar games. Or order *Starklight Random* today and play a new game every time you power up.

Sincerely,

*Aaron Smith*

Aaron Smith

Marketing Director

*P.S. Starklight Random comes with a 100% money-back guarantee—not that you'll ever need it!*

## E-Blast

Subject: Weekly News: Starklight Update

Vol. XXI Sept. 16, 2022

### *Starklight Random Release Moved Forward*

Just in time for holiday gift lists! *Starklight Random* will debut Nov. 2 with same-day shipping on all orders.

Read more>>

### *Q&A with Game Designer David Smith*

What inspired his best-selling game *Night Terror?* You may not want to know …

Read more>>

### *New Restrictions on Gaming Content?*

Four new bills in the US Congress seek to regulate game content. Should you be worried?

Read more>>

Unsubscribe or manage your communication preferences.


Credit: lassedesignen/shutterstock.com; ManuelfromMadrid/shutterstock.com; Marta Design/shutterstock.com; tanuha2001/shutterstock.com

# 4H Fundraising Messages and E-Blasts

## Purpose, Audience and Media

A fundraising message is an unsolicited message, often a business letter or email, sent to potential donors on a mailing list. Nonprofit organizations use fundraising messages to raise money, identify new donors, increase visibility, boost public relations, identify potential volunteers and publicize new programs. Your purpose in fundraising messages is to persuade recipients that a deserving cause needs their money and that you can spend it even better, with more impact, than they can on their own.

Fundraising messages rarely are single, one-time efforts. Successful fundraising generally requires a well-planned series of "touches," which are informative points of contact with potential donors.

Like traditional sales messages, fundraising messages require a highly targeted mailing list. Organizations can use their own in-house databases of donor and potential donor information. They also can rent highly targeted mailing lists from organizations known as list brokers. A typical response rate from a donor acquisition mailing is between 0.5% and 2.5%. However, the response rate from resolicitation—that is, seeking more money from a current donor—can be three to four times more successful than initial requests, particularly if your organization details exactly how it's spending the money.

Fundraising e-blasts are becoming more common, but experts say that the most successful fundraising emails remain individual messages sent from current donors to their friends. Several fundraising websites now allow you to automatically post to Facebook and other social media, informing friends and followers about your donation and encouraging them to join you. Organizations also can use their own social media posts to direct potential donors to fundraising landing pages (which should be optimized for easy phone access).

### KEY TO SUCCESS

A successful fundraising message delivers a personal, emotional, benefit-driven message directly to individual recipients. It shows them that they can make a difference. Effective, ongoing fundraising campaigns include frequent mailings and social media interactions with the receptive individuals.

## Format/Design

For fundraising letters, follow the general guidelines for the business-letter format on pages 297–301. However, like sales letters, fundraising letters often delete the recipient information from the heading. They often lack the three lines that include the recipient's name, address, city, state and ZIP code. They do include the date.

Like sales letters, fundraising letters often highlight key passages with design elements such as boldface type, different-colored type, underlining, capital letters, subheadlines—and even handwritten sticky notes, prepared and attached to the letter by a machine. However, fundraising letters shouldn't seem excessively expensive (an apparent waste of donors' money) or too flashy (inappropriately frivolous for an important social need).

E-blast fundraisers contain direct links to fundraising landing pages. Again, those sites should be optimized for easy mobile access.

## Content and Organization

A fundraising letter can be part of a direct-mail package (see pages 268–269). A fundraising package traditionally contains the following:

- An attention-grabbing outer envelope (perhaps with a scannable donation code, which might earn a postage discount—see Tip 3)
- A personalized fundraising letter
- A personalized reply form (digital donations lessen the need for this item)
- A reply envelope (digital donations lessen the need for this item)
- A brochure (optional)

Like a sales letter, a fundraising message traditionally consists of six sections.

### *Part 1 of 6: Begin With a Teaser Headline (Optional)*

In fundraising letters, teaser headlines are optional. A teaser headline appears in the upper-left corner of the page, generally above the date and the salutation. Teaser headlines usually use a different typeface—and often a different color—from the rest of the type. They are larger (usually 18- or 24-point type), and they don't extend across the entire page. Instead, they split into two or three lines and remain in the upper-left corner.

Unlike newspaper headlines, teaser headlines usually don't tell; instead, they tease. They ask a question, mention a problem, state an eye-popping statistic or refer to a solution. The goal of a teaser headline is to capture readers' attention and get them to read the letter in order to learn about the headline. Avoid any teasing strategies that seem frivolous or that seem to lessen the seriousness of the letter.

Fundraising e-blasts generally replace this headline with the email subject line—for example, "How can $2.98 save a life?" The subject line must be powerful enough to capture recipients' attention and entice them into opening the email. Subject lines of 40 or fewer characters and spaces can help ensure that recipients see the entire subject.

## Part 2 of 6: Present an Emotional Description of the Need

In the first paragraph or paragraphs, describe the social problem in specific, emotional terms. Grab readers' attention and help them identify with the cause or issue. Begin with a piece of genuine news or a real-life story that evokes empathy. Your goal is to show readers that something in life needs to be better. Use a personal salutation and, if possible, speak directly to readers within the letter, calling them by name. Avoid first names, however. Use a courtesy title—Mr. or Ms.—and the reader's last name.

This section can exceed one paragraph, but don't let it dominate the letter. Don't overwhelm readers with the scope of the problem. Show that the situation is serious—but not hopeless. At this point, keep the focus on the problem, not on the solution.

## Part 3 of 6: Present Your Organization as a Solution to the Problem

In a new paragraph, present your organization as a solution to the problem. Concentrate on illustrating the benefits created by the organization. Show how your organization's good works have impacted society.

Part 3 can be several paragraphs. As noted in the Format/Design section, this portion often features design elements such as boldface type, different-colored type, underlining and capital letters. Consider using such elements to highlight particular strengths of your organization and its successes.

## Part 4 of 6: Ask for a Donation

Wait until this section to ask for a donation. By this time you've built a case for why your organization can make a difference. Break the dollar amount into understandable figures—for example, "For only 50 cents a day you can provide a nutritious meal to a hungry child." Ask for a specific amount, or offer recipients several giving levels. Be sure to explain that the donation is tax deductible and that your organization can supply a receipt. Create a sense of urgency by giving readers a reason to respond now. Offer easy online ways to donate. A fundraising e-blast would include a link at this point.

### *Part 5 of 6: Re-evoke the Sense of Need or Desire and Again Ask for a Donation*

Close by re-evoking the sense of need. Remind readers once more of the problem and how giving to your organization is the solution. One person should sign the letter, preferably in blue ink to make it appear personal. Include a standard "Sincerely" sign-off.

### *Part 6 of 6: Add a P.S.*

Always include a P.S. It should reinforce your strongest reason for giving now. Place the P.S. below the sender's signature and title. (Include an extra space, just as if the P.S. were a new paragraph.) In letters, some postscripts even appear to be handwritten, as if they were an urgent personal note from the sender to the recipient.

A P.S. is not signed. It appears at the bottom of a message and usually is introduced simply by the initials *P.S.*

## FUNDRAISING TIPS

1. **Complete an SMP or creative brief**: Consider completing a strategic message planner or creative brief (pages 171–195) before you write your fundraising letter. These documents can help you focus on your recipient's interests and help you deliver one, clear, goal-oriented message. Review the discussions of sales messages (pages 265–272) and direct-mail packages (pages 268–269) for more information on writing successful fundraising messages.
2. **Use multiple channels**: Support your fundraising letter or e-blast with social media and news media outreach. MobileCause, a creator of fundraising software, found that recipients are 50% more likely to donate when they receive your message and learn of the need via several media.
3. **Study postal regulations**: For example, the United States Postal Service offers a 2% mailing discount for envelopes that include scannable codes that simplify donations from phones.
4. **Go visual**: In fundraising e-blasts and brochures, use powerful visual elements to attract recipients' attention and evoke an emotional response.

## Fundraising Letter

2010 Ridglea Dr. • Austin, TX 55111 • 555-999-5555

*Can a computer really save a kid's life?*

May 29, 2023

Dear Ms. Shakur:

Henry Smith, age 12, is what society calls a throw-away kid. His dad left before Henry was born. His mom died four years ago. Henry's grandmother is raising him. She does her best, but she can't keep up with an active sixth-grader. Henry has a criminal record—breaking and entering—and may be headed toward reform school. Six of his friends are there already. Henry says he doesn't want to follow them. Henry says he wants a chance, any chance, to turn things around.

Ms. Shakur, I think Henry has found that chance: It's an after-school program called **Tech for Texas Kids**—or just TTK. *TTK* provides computers and computer instruction for kids from kindergarten through 12th grade. TTK operates after school and during weekends. So instead of getting into trouble, Henry and his friends are learning, developing job skills and having fun, all under adult supervision. TTK operates in more than 1,000 public schools in Texas.

> *"My biggest interest in the world right now is computer graphics," Henry says. "When I grow up, I'm going to be a virtual-reality designer. I'm going to buy the nicest house you ever saw for my grandma."*

Tech for Texas Kids is sponsored by MGS Games Foundation of Austin, Texas. With your help, Ms. Shakur, we can buy enough computers for every precious child who wants to join TTK. Please visit our website at www.mgscck.org or use the scannable code on the outer envelope to donate $50, $75, $100 or more. It's tax-deductible, and we'll send you a receipt.

Henry Smith isn't a throw-away kid. Neither are the thousands of others who need Tech for Texas Kids. Please help us provide the computers that save kids' lives.

Sincerely,

*Mary Adams*

Mary Adams
Foundation Director

*P.S. Governor John Jones says Tech for Texas Kids is "the best program I've ever seen for turning at-risk children into great citizens."*

## E-Blast

Subject: Can a computer really save a kid's life?

Tech for Texas Kids provides after-school and weekend computer instruction and fun to more than 40,000 at-risk children every year.

Governor John Jones calls it "the best program I've ever seen for turning at-risk kids into great citizens."

Your donation of $50, $75, $100 or more can help us reach even more precious children. Give today and save a life!

*Tech for Texas Kids is sponsored by*

Unsubscribe or manage your communication preferences.



Credit: Pressmaster/shutterstock.com; tanuha2001/shutterstock.com

# 41 Brochures

## Purpose, Audience and Media

Brochures, booklets and fliers are strategic communication messages printed and distributed to a specific audience for a specific purpose. The three differ slightly in their formats. Brochures are typically a single piece of paper printed on both sides and folded into panels. They can use full color, spot color (accent color) or black ink only. Booklets, on the other hand, are printed in four-page increments and usually are saddle stitched (stapled down the middle, forming a mini-book). They typically are full color and are ideal for company annual reports or elaborate promotional pieces. Fliers are mini-posters, printed on one side of the page and intended for bulletin boards or hand delivery. They usually are quick-copied on colored paper. They are ideal for event announcements and one-page advertisements. Although this section will focus on writing and designing brochures, the basic principles also apply to booklets.

Brochures inform or persuade. They usually are part of a marketing plan media mix that might include print ads, video ads, audio ads, direct-mail packages, social media posts or billboards. Because of their abbreviated length, brochures don't tell the whole story; they merely deliver highlights. They can be used as material on display racks, enclosures in direct-mail packages, handouts, leave-behind sales materials or stand-alone direct mailers. They might advertise a product, recruit volunteers, make people aware of an issue or announce a workshop, lecture, performance or conference.

When defining your target audience, answer these questions in addition to the typical demographic, psychographic and behavioral queries:

- How will your audience receive the brochure?
- Why is your audience reading this brochure?
- What does the audience already know about your product?
- What is the audience's current attitude (if any) toward your product?
- Where else will the audience encounter the message of the brochure?

As noted, brochures exist primarily on paper, though sales forces sometimes use brochures on tablets such as iPads and as downloadable PDFs.

**KEY TO SUCCESS**

Effective brochures marry words and images to deliver a single message to a specified audience.

## Content and Organization

Like any effective advertising message, a brochure begins with the completion of a strategic message planner or creative brief (pages 171–195). Those documents help to define the audience and focus the message. The target audience determines the tone, vocabulary and type of appeal the message takes. The strategic message must be clear and specific. A brochure designed to speak to several audiences or accomplish several goals is doomed to fail. For example, a brochure designed to recruit volunteers for a hospital auxiliary won't work to invite new community members to use the hospital's outpatient services.

Brochures need a theme or unifying concept that amplifies the strategic message. The theme helps to interpret, define and present the message to the audience. The theme might take the familiar and give it a new twist. It might create a visual metaphor and show how two seemingly unrelated topics share many characteristics. It might create a distinct personality for the printed piece. For example, a brochure on financial services might be illustrated with images of a needle, thread and tape measure, all tools used by a tailor. The implication is that the same attention and personal service needed to make a custom suit will be given to building a client's financial portfolio. A brochure on pediatric services at a hospital might be illustrated with crayon drawings, giving it a feeling of youthful exuberance.

Sometimes themes are created with words, and other times they are created with graphic images. Often, both words and images develop a theme. A brochure might begin and end by telling a story. Another brochure might repeat graphics to link elements together. Both provide a unifying element.

### *Panels*

Because brochures are typically folded into panels, the designer must create a clear roadmap to direct the reader into and through the piece. The size, shape and number of folds determine how the reader views the piece. The six-panel brochure, also called a tri-fold brochure, is the most common organizational structure (see Figure 4.1). Other folds and designs are discussed later in this section. Subheadings help keep passages short and keep the reader focused.

| Panel 2 | Panel 6 | Cover |
|---|---|---|

| Panel 3 | Panel 4 | Panel 5 |
|---|---|---|

**FIGURE 4.1**

In the six-panel brochure, the *front cover* invites the reader into the piece. It catches the eye and provides a visual focus. It hooks the reader with a provocative headline, a question or a compelling image. It may contain a teaser headline with a subheading that explains the nature of the piece. Usually the cover contains an image, a headline and the organization name and/or logo.

*Panel 2* is the next most-likely panel to be read because of its position in the six-panel format. (You may understand this better if you take a sheet of paper, mark the front and back just as in Figure 4.1 and then fold the sheet.) Panel 2 typically presents a stand-alone message that summarizes the reason the customer should choose this product. It also may reinforce key points, begin a compelling story or present a testimonial. It often is written after the main copy message.

*Panels 3, 4* and *5* present the main copy message and are viewed as one three-column unit. This copy clearly explains the product's features and benefits. The message has a distinct beginning, middle and end, much like an essay. This three-panel section often includes subheadings.

*Panel 6* is the back cover. This is the panel people are least likely to read, so avoid continuing the copy message to this space. Use it for contact information such as the telephone/fax number, website address, email contacts and physical address. It often repeats the organization name and logo. It also can be used for a recipient's address if you design the brochure to be mailed.

As mentioned previously, brochure copy—particularly in the cover plus panels 3, 4 and 5—has a beginning, middle and end. Here are some copy approaches that help organize and deliver the message. Choose the one that best fits your audience.

## The Beginning

- **Ask a question**: Questions invite conversation and break down barriers. They imply that an answer is forthcoming and that reading further might provide that answer.
- **Pose a problem**: Position your brochure as the answer to a problem. If the reader shares that problem or is interested in it, they will read further.
- **Offer an opportunity**: Many people can't resist the chance to experience new things, whether they are new products or new experiences.

- **Set a mood or create an emotion**: Emotions are powerful persuaders. Evoking nostalgia or empathy helps the reader identify with your cause.
- **Tell a story**: Hook your reader with an intriguing story. Get them to identify with a character or action.

### *The Middle*

- **Make key selling points**: Describe your company's or organization's competitive advantages. Write in terms of features and benefits. Make one point per subheading.
- **Give the solution**: If you presented a problem in the opening, now is the time to explain the solution.
- **Arrange information from least important to most important**: Sometimes it works best to save your strongest arguments for the end and build your argument gradually. Test different organizational strategies on sample, representative audiences if you're not certain about the best organization.
- **Explain the steps in a process**: If your information is sequential or requires elaborate explanations, break it down into manageable bits. Explain each step under a separate subheading.

### *The End*

- **Summarize the main points**: Repeat your key points. Leave the reader with a clear idea of why they should act now.
- **Remind the reader of the importance of the topic**: Make one more appeal for why your topic deserves consideration.
- **Link the end with the story in the beginning**: If you began by telling a story, finish it or give some indication of how it ends. The reader needs closure. Don't leave them hanging.
- **Make a call to action**: Tell the readers what they should do now. Leave no doubt in their minds what their next action should be: Call, give, buy, volunteer or return a coupon.

## Format/Design

Good design provides the structure and form that hold the piece together. Good design works seamlessly with the copy to reinforce the message. A well-designed publication:

- Enhances readability
- Amplifies the message

- Organizes the message
- Is practical
- Doesn't call attention to the design alone

The adage that you have only one chance to make a good first impression especially holds true for brochures. If a reader isn't attracted to the cover, chances are they won't pick up your brochure.

In designing a brochure, you must select the format, type, layout, color and paper.

## *Format*

One of the first considerations in designing a brochure is deciding what size it needs to be. Ask yourself the following:

- How will the piece be used?
- Does it need a return coupon?
- Should it be vertical or horizontal?
- How many panels will it need?
- How will it be distributed?
- Will it be mailed?

The standard six-panel brochure can be folded from an 8.5-by-11-inch piece of paper and easily fit into a #10 business envelope. If cost is a consideration, avoid running images to the edges, since that requires trimming an oversized sheet and adds expense. If cost is not your primary consideration, use a 9-by-12-inch piece of paper. It still will fit into a #10 envelope but gives you more space for your message. If your piece is a self-mailer, consult your local post office for standard mailing sizes and rates. Odd-sized pieces require additional postage. It's best to know before the piece is printed that it will cost more to mail it. It's always a good idea to talk with your local printer throughout the design stage, especially if you are creating an unusual piece. They can often make suggestions that will save you money and time.

## *Type*

A good basic rule is to use no more than two typefaces in a publication. For example, you might use Franklin Gothic (a sans-serif typeface) for headlines and Garamond (a serif typeface) for body copy. These typefaces provide good contrast in styles. Use consistent type styles for headlines and subheadings. Avoid using tilted headlines or vertical type. Headlines set in all caps are difficult to read, so use them only for very short headlines. Likewise, italic type is difficult to read. Use it sparingly.

## Layout

Keep your layout clean and balanced. Use the basic principles of design: balance, movement, emphasis, contrast, proportion, space and unity (see pages 35–41). Use negative space (white space) to your advantage, and don't make your layout feel cluttered or too busy. The 70–30 rule says use 70% text and 30% white space. Use generous leading (the space between lines of type—pronounced *ledding*) to enhance readability. Use short line lengths to make copy easier to read. And remember: Be consistent with your layout throughout your brochure.

## Color

Color definitely enhances design. A full-color brochure printed on glossy paper jumps out and demands attention. Color affects the design and production process from the beginning, so make this decision carefully and early (see pages 38–41). If your budget won't permit full color, you can use spot color or two-color printing. Use the additional color for headlines, subheadings and graphics. Don't overdo the accent color, and consider carefully before you use it on photographs. Coloring people blue in photographs seldom works.

## Paper

Paper adds texture to a publication. It makes it a three-dimensional experience. Always work with your printer when choosing paper. They often have a house stock that will save you money. Ink performs differently on different papers, so ask lots of questions.

When choosing paper, consider the finish, weight and color. Paper is either coated or uncoated. Coated paper has a hard, enamel finish and comes in gloss, matte and dull. It is ideal for full-color pieces. Uncoated paper has a smooth, vellum or pattern finish, like linen or laid.

Paper is categorized by weight. It comes in text weight and cover weight. Either will work for a brochure. If you use cover weight, you may need to have the folds of your brochure scored to prevent cracking and tearing along the edges (see Figure 4.2). Also, if you are mailing your piece, check with the post office concerning the weight restrictions.

Includes numerous panels that are unrolled slowly to continue the message.

Paper comes in thousands of shades and colors. As a general rule, full-color photography works best on white or cream paper. An inexpensive way to add color to your brochure is to choose a colored paper stock and print in black ink.

Figures 4.3 and 4.4 show examples of a six-panel brochure. Read the copy and study the layout. Study how the words and design work together.

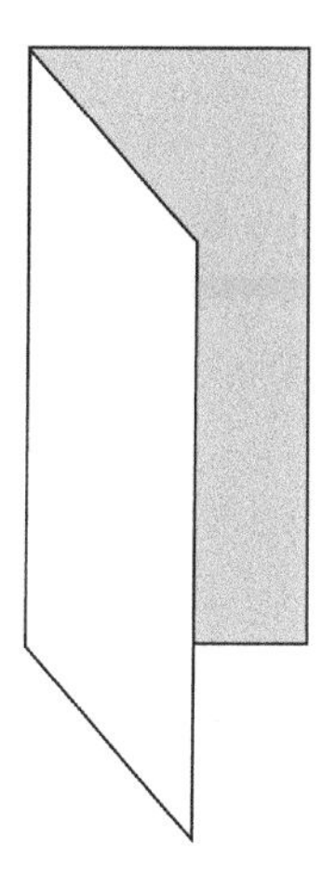

**(a) Parallel fold**

Creates four panels that can be used vertically or horizontally.

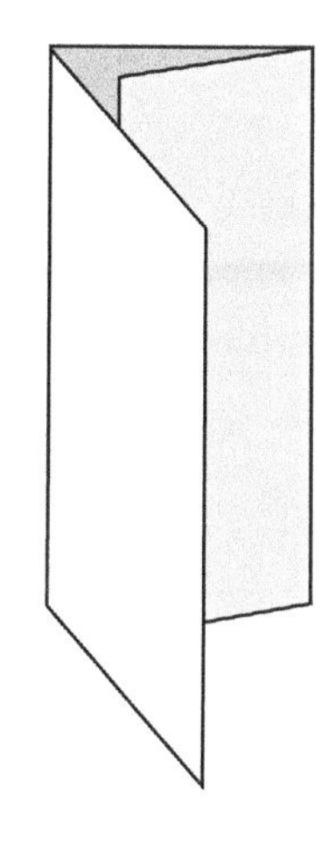

**(b) Letter fold**

Creates six panels that can be used vertically or horizontally.

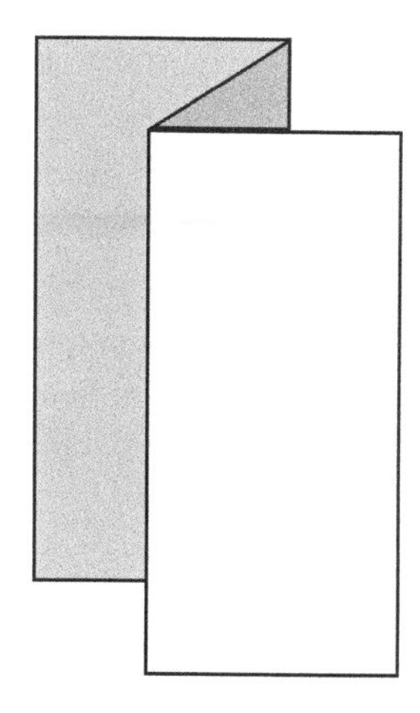

**(c) Accordion fold**

Creates six panels; the reader views one side at a time.

**(d) Gate fold**

Creates panels that open like doors; the reader views the inside information at one time.

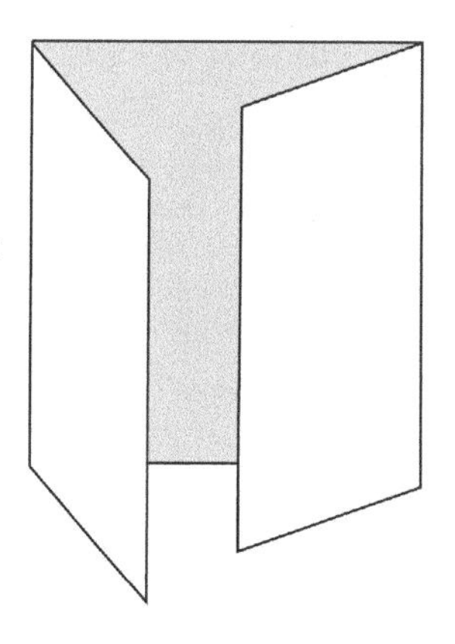

**(e) French fold**

Creates a poster when completely opened and is ideal for large layouts.

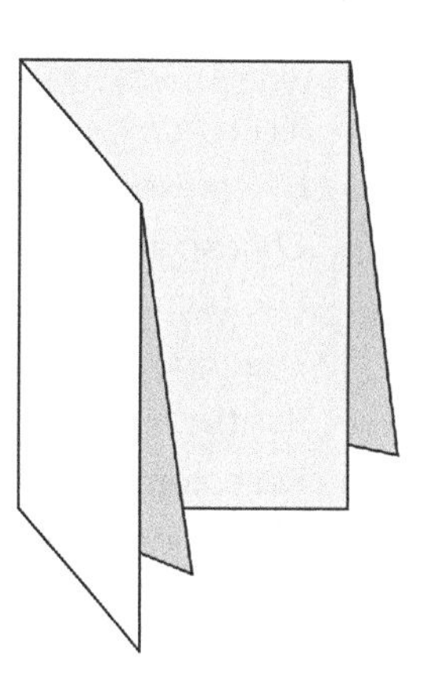

**(f) Double parallel fold**

Creates eight panels; the first fold creates a double-page spread. The second fold opens to four full panels. This fold is ideal if the brochure needs a coupon or return card.

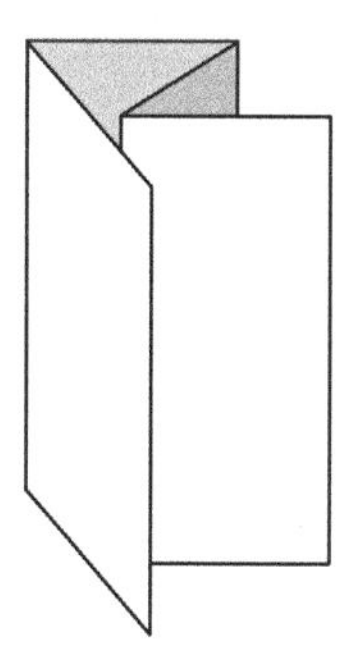

**(g) Barrel fold**

Includes numerous panels that are unrolled slowly to continue the message.

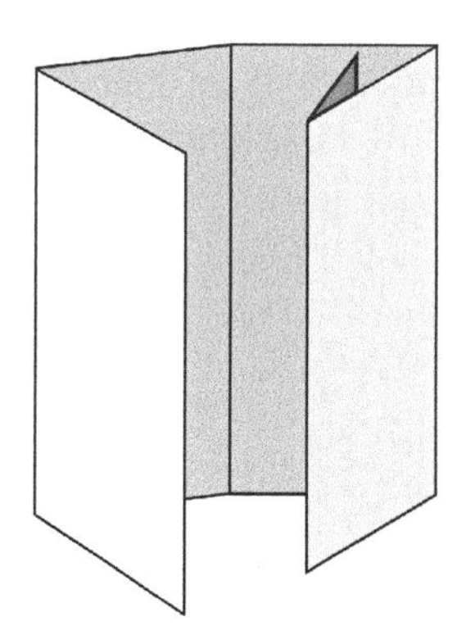

**FIGURE 4.2**

## TIPS FOR BROCHURE COPY

1. **Use strong headlines and subheadings that lead the reader into the text**: These headings should make the reader curious or make them think they will learn something if they continue. A benefit-driven headline might be "How to grow award-winning roses."
2. **Create good headlines that**:
   - Grab attention
   - Set the tone
   - Carry much of the message
   - Have a visual complement that supports the message
3. **Use subheadings to break up information into manageable chunks**: Let the reader feel that they can read any section independently of the other sections. Subheadings should flow from the main headline. Collectively, they should describe the page content. Label-style headlines work well for subheadings. For a product, label-style headlines could be single words such as *Colors*, *Sizes* and *Prices*.
4. **Speak directly to the reader in a casual, informal tone**: Remember, *you* is the most important word in persuasion.
5. **Use present tense and active voice** (pages 11–12).
6. **Choose a tone appropriate to your audience**.
7. **Use bullets to list information**.
8. **Use parallel construction**: See page 348.
9. **Put the emphasis on what the reader will gain**: What's in it for me?
10. **Keep copy short**: Use short sentences of 15 words or less. Use short paragraphs of no more than 15 lines. Use sentence fragments and phrases if they are appropriate for the rhythm of the copy.
11. **Use details, details and more details**: Don't write in generalities. You must know your product or service thoroughly and tell the reader the details of its benefits. Refer to the SMP or creative brief.
12. **Use imperative mood, just as this sentence does**: Command your reader to action.
13. **Choose short words rather than long words**: For example:
    - *Achievement* can be replaced with *success*.
    - *Advantageous* can be replaced with *good* or *cheap*.
    - *Utilize* can be replaced with *use*.
    - *Employment* can be replaced with *work*.
14. **Avoid puffery**: See page 57. Don't be melodramatic. Avoid clichés, buzzwords and unfamiliar acronyms. Make valid claims. Don't exaggerate your product or service. Double-check your facts.

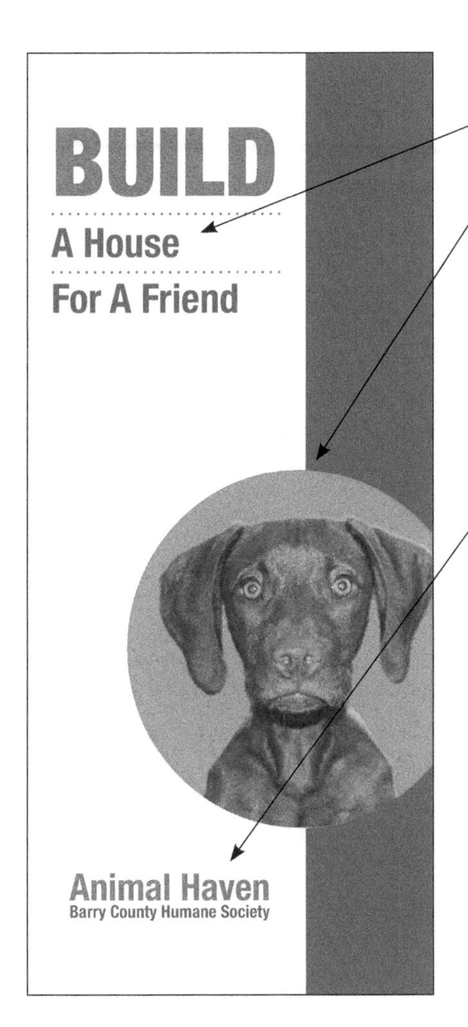


The cover invites the reader into the piece.

The headline speaks directly to the reader and calls them to action.

The image adds interest and reinforces the headline.

Graphics build the theme using the circle design that mirrors the logo (see panel 6). Dots are used as a graphic element throughout the piece. The vertical bar creates contrast with the circle graphic and adds asymmetrical balance (see page 36).

The name of the organization appears on the cover.

**Animal Haven ... A Community Resource**

Each year we care for more than 7,000 animal friends and serve the animal-related needs of approximately 100,000 people. The average cost of care for each animal is $450 during its stay.

As a nonprofit organization, Animal Haven relies on a Barry County government grant and on donations from you, our generous supporters.

**Mission Statement:** Animal Haven acts as an advocate for all animals whether they are companion animals, wildlife, farm animals or animals in laboratories. We believe in the humane treatment of all living creatures. We believe that through compassion and caring we can make a difference in the lives of animals and in the lives of the people who love them.

The image connects people and pets in an emotional way that encourages nostalgia.

Panel 2 stands alone. The copy introduces the organization and explains its purpose. The mission statement adds credibility.

**FIGURE 4.3**

### Animal Haven ... A Place For Love

There is a place in our community where the hungry are fed, the homeless are sheltered, the abused are treated and the lost are returned. That place is Animal Haven, the Barry County Humane Society.

### Shelter And Adoption Services

As an Open Admissions shelter, we accept every animal that is brought to us, regardless of age, health, behavior or special needs. Our number one goal is finding responsible, caring homes for animals.

Whether the animal is lost or abandoned, we find joy in uniting people and pets. Last year alone, 675 dogs and cats were returned to their owners and 1,546 were adopted.

We offer special programs to help you choose the right pet for your family. The initial adoption fee for dogs, cats and rabbits of $250 includes

- Professional health examination
- First vaccinations
- De-worming (for dogs and cats 4 months old)
- Microchip implant
- Access to an animal behaviorist
- Educational materials on the care of the animal
- Spaying/neutering at Animal Haven

Panel 3 begins the copy message with an opening that sets the tone. It metaphorically compares Animal Haven to relief programs for people.

Copy blocks are broken into manageable chunks under separate subheadings. Notice the specific details included in the copy.

Bullets are used to list points.

### Vaccination And Health Services

Animal Haven provides health care for companion animals, sterilization surgery for dogs, cats and rabbits and microchip services. We have four licensed veterinarians on staff who care for animals of all ages.

Open Monday-Friday, 9 am to 7 pm, the clinic provides vaccinations, including rabies, distemper and bordatella, as well as de-worming, heartworm prevention, flea and tick control and laboratory testing services. We also provide emergency care for injured, lost and stray animals.

For a nominal fee, we implant a microchip in your pet that will help more than 12,000 shelters, like Animal Haven, identify you as the owner. There's no better way to instantly reunite you and your pet.

To help low-income families care for their pets, we offer the Prevent a Litter Assistance Program that subsidizes spaying or neutering services. The CARE discount program also helps families pay for emergency medical treatments.

### Community Education Programs

Animals teach all of us about compassion, empathy and unconditional love. Animal Haven brings animals and animal issues into the lives of schoolchildren throughout the county. Staff and volunteers lead discussions about pet care, local wildlife and other animal-related topics through

- Classroom visits with dogs, guinea pigs, rabbits and other living creatures
- Dogtalk, attention-grabbing, interactive presentations about dog behavior and safety
- Reptiletalk, discussions regarding the ethical and practical issues of keeping reptiles as pets
- Operation Pet Pals, a week-long curriculum designed for children K-6

Panel 4 continues the message. Two more subheadings explain key features as benefits.

Bullets are used to list points.

**Help Us Build A New Haven**

Founded in 1964, Animal Haven began as a mission to foster humane treatment of farm animals. The society purchased four acres of land and constructed a shelter on New Hope Road. Now 50-plus years later, we've outgrown our present facilities and plan to build a new complex on the adjoining six acres.

But building a new house for a friend won't be possible without your support. Help us continue to provide community education and animal rescue programs for Barry County. Help us build a safe haven for all creatures great and small. Help us be a place for love.

RSVP by sending your donation to:

Animal Haven
435 New Hope Road
Anywhere, USA 12345
(444) 777-2345

Closing copy provides more detail about the organization.

The final paragraph makes a call to action and links the closing with the opening copy theme.

The final call to action gives specific information about where to send a donation.

**How You Can Help Animal Haven**

We rely on support from generous donors, like you, to provide comfort and care for unwanted animals. Here are ways you can help.

- Give to the Build a House for a Friend Fund.
- Donate food, old blankets, towels or other needed supplies.
- Volunteer your time. Become a pet caregiver. Bathe and groom the animals. Walk dogs or play with cats. Stuff envelopes for mailings or help publicize an event.
- Find a special friend. Choose your next pet from Animal Haven and give a dog or cat a good home.
- Be a responsible pet owner. Keep current identification on your dog or cat. Keep vaccinations current and have your pet spayed or neutered.
- Teach your children to be humane citizens. Set an example by treating all animals with compassion and kindness.

**Animal Haven**

435 New Hope Road
Anywhere, USA 12345
(444) 777-2345
FAX (444) 777-3456
www.animalhaven.org

**FIGURE 4.4** Panel 6 stands alone. Since this panel is the least likely to be read, nonessential information is placed here. Bullets are used to list points.

Logo and contact information appear at the bottom of panel 6.

# Brochure

### Animal Haven … A Community Resource

Each year we care for more than 7,000 animal friends and serve the animal-related needs of approximately 100,000 people. The average cost of care for each animal is $450 during its stay.

As a nonprofit organization, Animal Haven relies on a Barry County government grant and on donations from you, our generous supporters.

**Mission Statement:** Animal Haven acts as an advocate for all animals whether they are companion animals, wildlife, farm animals or animals in laboratories. We believe in the humane treatment of all living creatures. We believe that through compassion and caring we can make a difference in the lives of animals and in the lives of the people who love them.

### How You Can Help Animal Haven

We rely on support from generous donors, like you, to provide comfort and care for unwanted animals. Here are ways you can help.

- Give to the Build a House for a Friend Fund.
- Donate food, old blankets, towels or other needed supplies.
- Volunteer your time. Become a pet caregiver. Bathe and groom the animals. Walk dogs or play with cats. Stuff envelopes for mailings or help publicize an event.
- Find a special friend. Choose your next pet from Animal Haven and give a dog or cat a good home.
- Be a responsible pet owner. Keep current identification on your dog or cat. Keep vaccinations current and have your pet spayed or neutered.
- Teach your children to be humane citizens. Set an example by treating all animals with compassion and kindness.

**435 New Hope Road**
**Anywhere, USA 12345**
**(444) 777-2345**
**FAX (444) 777-3456**
**www.animalhaven.org**

## BUILD
## A House
## For A Friend

**Animal Haven**
**Barry County Humane Society**

### Animal Haven … A Place For Love

There is a place in our community where the hungry are fed, the homeless are sheltered, the abused are treated and the lost are returned. That place is Animal Haven, the Barry County Humane Society.

### Shelter And Adoption Services

As an Open Admissions shelter, we accept every animal that is brought to us, regardless of age, health, behavior or special needs. Our number one goal is finding responsible, caring homes for animals.

Whether the animal is lost or abandoned, we find joy in uniting people and pets. Last year alone, 675 dogs and cats were returned to their owners and 1,546 were adopted.

We offer special programs to help you choose the right pet for your family. The initial adoption fee for dogs, cats and rabbits of $250 includes

- Professional health examination
- First vaccinations
- De-worming (for dogs and cats 4 months old)
- Microchip implant
- Access to an animal behaviorist
- Educational materials on the care of the animal
- Spaying/neutering at Animal Haven

### Vaccination And Health Services

Animal Haven provides health care for companion animals, sterilization surgery for dogs, cats and rabbits and microchip services. We have four licensed veterinarians on staff who care for animals of all ages.

Open Monday-Friday, 9 am to 7 pm, the clinic provides vaccinations, including rabies, distemper and bordatella, as well as de-worming, heartworm prevention, flea and tick control and laboratory testing services. We also provide emergency care for injured, lost and stray animals.

For a nominal fee, we implant a microchip in your pet that will help more than 12,000 shelters, like Animal Haven, identify you as the owner. There's no better way to instantly reunite you and your pet.

To help low-income families care for their pets, we offer the Prevent a Litter Assistance Program that subsidizes spaying or neutering services. The CARE discount program also helps families pay for emergency medical treatments.

### Community Education Programs

Animals teach all of us about compassion, empathy and unconditional love. Animal Haven brings animals and animal issues into the lives of schoolchildren throughout the county. Staff and volunteers lead discussions about pet care, local wildlife and other animal-related topics through

- Classroom visits with dogs, guinea pigs, rabbits and other living creatures
- Dogtalk, attention-grabbing, interactive presentations about dog behavior and safety
- Reptiletalk, discussions regarding the ethical and practical issues of keeping reptiles as pets
- Operation Pet Pals, a week-long curriculum designed for children K-6

### Help Us Build A New Haven

Founded in 1964, Animal Haven began as a mission to foster humane treatment of farm animals. The society purchased four acres of land and constructed a shelter on New Hope Road. Now 50-plus years later, we've outgrown our present facilities and plan to build a new complex on the adjoining six acres.

But building a new house for a friend won't be possible without your support. Help us continue to provide community education and animal rescue programs for Barry County. Help us build a safe haven for all creatures great and small. Help us be a place for love.

RSVP by sending your donation to:

Animal Haven
435 New Hope Road
Anywhere, USA 12345
(444) 777-2345

**FIGURE 4.4**

SECTION

# 5

# Strategic Writing in Business Communication

## OBJECTIVES

**In Section 5: Strategic Writing in Business Communication, you will learn to write these documents:**

- Good-news correspondence
- Bad-news correspondence
- Job-request correspondence
- Résumés
- Memoranda
- Business reports

# 5A Introduction to Business Communication

Business communication is the exchange of messages that help an organization complete its day-to-day functions. Business communication is sometimes called administrative communication because it helps an organization manage basic routines.

In a moment, we'll define business communication—and, more precisely, business writing—by the documents it includes. First, however, we might better understand business communication if we compare it to family communication. Let's imagine a mythical family consisting of Mom, Dad, 2.5 kids, a dog and a goldfish. Think of all the communication required just to help that family function smoothly every day. Family members discuss what time dinner will be and when soccer practice is. Mom tells the kids to clean their rooms; the kids explain why they can't right now. Dad asks if anyone has fed the dog. The kids make suggestions for summer vacation. Both parents ask how the homework is going, and the kids ask for more allowance money. To keep the family functioning, its members also discuss family business with outside groups: neighbors, teachers, babysitters and others.

All that communication holds the family together. It helps family members sort out their priorities and schedules; it helps them plan, debate and establish policies that members will follow. And that's a lot like business communication. Employees of an organization are like family members, and business communication is the exchange of messages that allow the group to function effectively.

The most effective method of business communication is often face-to-face conversation. Studies conducted by the International Association of Business Communicators show that face-to-face communication with the boss is usually an employee's favorite way to learn important news about the organization.

However, much of business communication needs to be written. Writing allows more than one person to see the same message. Writing can create a permanence that face-to-face communication can't match. As effective as face-to-face communication can be, organizations rely on good, clear, strategic writing to function from day to day.

Written communication also can help counter the effects of an unreliable channel of business communication: the grapevine. Many of us love to gossip, and offices supply ample opportunities for rumors and false stories. Good, frequent business communication can take the air out of rumors.

The documents in this section help define business communication: business correspondence, memoranda and reports. This section also treats *you* as a business and includes two documents that are very important to your future: job-request correspondence and résumés.

Whatever your career interest is—public relations, advertising, sales, marketing or something else altogether—chances are strong that your dream job will include many aspects of business communication.

# 5B Social Media in Business Communication

## Purpose, Audience and Media

The uses of social media in business communication tend to cluster into three areas: communication with potential employees, internal communication and employees' use of private social media in the workplace. A fourth, related area involves regulating the use of social media within an organization. Audiences for social media within this category of strategic writing generally are fellow employees or business associates, such as suppliers or distributors.

## Communication With Potential Employees

For college students seeking a job or internship, social media can be a powerful help—or a powerful hindrance. On the positive side, LinkedIn allows you to present your own credentials and to follow, study and connect with potential employers who can use the same platform to reach out to you. If you're new to LinkedIn, the Help Center of that popular social media platform offers advice on creating an effective profile and building connections to potential references and employers. On the negative side, surveys show that at least two-thirds of all employers will, as much as possible, check out your social media presence to assess your personality and reliability. Tagging and sharing party photos may be fun for a few months, but the cost of such documentation might include a lack of job offers.

## Internal Communication

Programs such as Yammer, Slack and Trello offer project-management features that allow colleagues to simultaneously create, organize, share and edit specific documents and larger communication campaigns. Slack, for example, has the slogan "Where Work Happens," and its features include direct messaging, voice and video calls, file sharing, archiving and much more. Yammer's slogan is "Connect and Engage Across Your Organization," and Trello assures users that its functions will allow them to "work more collaboratively and get more done." Those programs of course have apps for tablets and phones. Programs such as Periscope and Facebook Live are increasingly popular for training and demonstration purposes. Usage of these programs can extend beyond an internal audience, of course, allowing an organization to collaborate with vendors or other business partners.

## Private Social Media in the Workplace

Believe it or not, your authors recommend avoiding writing too much in this area—that is, we recommend curtailing your personal social media activities during working hours. As early as 2016, the Pew Research Center reported that almost 80% of U.S. workers confessed to using personal social media accounts during working hours. The top three reasons were taking a mental break from work, touching base with friends or family and professional networking. The fourth reason, at least, was gathering information for work. Research shows that most employers frown on non-job-related use of social media during working hours: A hot legal issue today is whether organizations can fire employees for violations of company social media policies. (The answer? It depends on the nature of the business and the terms of your employment.) However, research also shows that taking short mental breaks can increase workplace productivity and that employees' positive, personal messages about their organization move faster and farther than many official organizational communications—which leads to our next section.

## Workplace Social Media Policies

Organizations that have found beneficial ways to address private social media use during working hours tend to have a similar approach: They ask their employees to help them create realistic policies. Best practices for these policies include:

- Allowing brief social media breaks throughout the day
- Reminding employees that they represent their employer even in their private social media activities
- Forbidding employees from sharing private organizational news
- Asking employees to alert their employers if they become involved in heated, threatening social media exchanges
- Reviewing and updating the policy at least twice a year

In the United States, the National Labor Relations Board publishes periodic updates on legal guidelines for regulating employees' use of social media during working hours.

If trends in freelancing, flexible work hours, telecommuting (or working from home) and global business relationships continue, the presence of social media within business communication no doubt will increase—and that's good news for strategic writers.

# 5C Business Correspondence

## Purpose, Audience and Media

Business correspondence generally takes the form of emails or paper letters. Both forms deliver strategic messages to targeted individuals, and both are concise: The business letter, for example, is usually one page, plus an envelope. There are several different kinds of business correspondence. Among the most common are:

- Good-news correspondence (pages 302–305)
- Bad-news correspondence (pages 306–310)
- Sales correspondence (pages 265–272)
- Pitch correspondence (pages 118–123)
- Job-request correspondence (pages 311–316)

The audience of a business letter or email usually is one person. Your organization generally wants something from that individual. Understanding the values and self-interests of the recipient of a business message increases your chances of getting what you want.

The format of a business letter generally is one sheet of your organization's stationery that you'll put in an envelope and place in the mail. Email messages have mostly replaced slower, more expensive paper letters, but for many recipients, email still lacks the formality and gravity of an old-fashioned paper letter. Your knowledge of the intended recipient can help guide your choice of letter versus email.

This first segment on business correspondence presents the standard format for letters and emails. This section also includes content and organization suggestions that apply to most business correspondence, particularly letters, which can be unfamiliar territory for many of us.

**KEY TO SUCCESS**

Different business messages have different strategic organizations. For example, a bad-news message is organized differently from a sales message. The different ways to organize business messages support the strategic purposes of the different forms of business correspondence.

## Format/Design

With rare exceptions, keep a business letter to one page. Sales letters can be an exception to this guideline.

Whenever possible, use your organization's stationery. Use your organization's stationery for the envelope as well.

Single-space the lines of a letter. Include an extra blank line between:

- The date and the other heading (recipient) information
- The recipient information and the "Dear Mr." or "Dear Ms."
- The paragraphs of the letter (don't indent the paragraphs)
- The text of the letter and the "Sincerely"
- Your typed title at the bottom and any extra notes, such as "P.S." (for postscript) or "encl." (for enclosure)

Align the proper headings (date and recipient information) along the left margin. Figure 5.1 shows sample headings.

Type "Sincerely" at the bottom left. Because "Sincerely" is traditional, that word is almost always the best sign-off. "Sincerely" is courteous and conservative. A less traditional sign-off might draw attention away from your name and title.

After "Sincerely," skip down four to six spaces (enough room for a *legible* signature), and then type your first and last names. Under your name, type your title. Like the rest of the letter, these two lines should be aligned along the left margin. Don't type your organization's name under your title. Your organization's name is already on the stationery.

Remember to sign your name above your typed first and last names.

Notes at the bottom, below your typed title, are also aligned along the left margin:

- Adding *cc* means you've sent a copy to the person you name (for example, "cc: Mary Jones"). Because *cc* stands for "carbon copy" and is obsolete, you'll sometimes see just *c* for copy.
- The abbreviation *encl.* stands for "enclosure" and means that you've enclosed another document, such as a brochure, in the envelope with your business letter. (Be sure that you remember to enclose the additional document.)
- *P.S.* means postscript and is a brief extra note that you add to a letter after you've signed it. A P.S. can be an important element of a sales letter.

Figure 5.1 shows a sample closing for a business letter.

2010 Ridglea Dr. • Austin, TX 55111 • 555-999-5555

Today's date

Mr. or Ms. Recipient's First and Last Names
Recipient's Business Title
Recipient's Organization
Organization's Street Address
Organization's City, State (no comma) ZIP

Dear Mr. or Ms. Last Name:

Thank you very much for . . .

*When you're writing a letter that isn't on your organization's stationery, such as a job-request letter, add your own street address and city, state and ZIP code to the date section, which is above the recipient information. Note that you don't type your name in this section. Your typed name goes at the bottom. In other words, at the top of the letter you would type*

Your Street Address
Your City, State (no comma) ZIP
Today's date

. . . .

Again, thank you very much for your time and effort. I look forward to meeting you next Friday.

Sincerely,

*Your Legible Signature*

Your Typed First and Last Names
Your Title

cc: Mary Jones (this is optional)

*(If this is a personal letter, such as a job-request letter written by a college student, don't include a title below your typed name—you don't have a title yet.)*

**FIGURE 5.1**

## Content and Organization

Avoid an unintentionally sexist greeting. For example, if you're writing to Lynn Jones, is that individual male? Or is Lynn Jones female? Should your salutation be "Dear Mr. Jones" or "Dear Ms. Jones"? Find out the answer.

Don't use "Miss" or "Mrs." in salutations to female recipients unless you are certain that is the person's preference. Ordinarily, use "Ms." for all women unless a title such as "Dr." would be more appropriate.

When you address the recipient by a courtesy title ("Mr." or "Ms.") and a last name, place a colon (not a less formal comma) after the last name. With such a formal beginning, be sure to sign both your first and last names at the bottom of the letter above your typed first and last names and title (see the example on page 310).

If you know the recipient well and your salutation is simply "Dear Lynn," then place a comma (not the more formal colon) after the recipient's first name. With such an informal beginning, be sure to sign *just your first name* at the bottom of the letter above your typed first and last names and title (see the example on page 299).

In email messages, the subject line is important. Usually, the subject line concisely summarizes the email's main point: "Transportation services request." If a quick action or response is necessary, the subject line can indicate that: "Brochure photos—Approval needed." A subject line for a good-news message to someone you know well may simply be "Congratulations!" Subject lines for bad-news messages generally are neutral: "Response to your request."

Emails do not have headings similar to business letters. The first line of an email is the salutation: "Dear Ms. Jones" followed by a colon (in formal messages) or a comma (in informal messages).

The sign-off for an email often is "Sincerely" or "Thank you," followed by your name on a separate line. You can program your email software to automatically add your title and contact information to the bottom of the email.

Paragraphs in email messages resemble those in business letters: They are not indented. Instead, insert a blank line between paragraphs.

## BUSINESS CORRESPONDENCE TIPS

1. **Get to the point**: Within the first few sentences of most business messages, the recipient should know why they are reading the letter or email. In other words, tell them why you're writing. Sales messages and bad-news messages can be exceptions to this guideline.
2. **Give instructions**: The closing of a business message (the last or next-to-last paragraph) often specifies or suggests what the next action in the particular situation should be. When appropriate, this closing should say what you'll do next or what you hope the recipient will do next, or both.
3. **Be courteous**: Whenever possible, be polite. Opening and closing with a "thank you," whenever appropriate, is both good manners and good business.
4. **Avoid clichés**: Write like the warm, intelligent human being you are—not like a cold, unfeeling business machine. Avoid clichés such as "It has come to my attention" and "I regret to inform you." Such sentences are so overused that they sound insincere.
5. **Sign legibly**: An illegible signature in a business letter can suggest carelessness or an oversized ego, which are not good qualities in a healthy business relationship.
6. **Don't hand-letter an envelope**: The envelope provides the first impression. Use a good laser printer. The envelope should look as professional as your letter itself. In the center of the envelope, devote four lines to the recipient's name, the name of the organization, the street address, and the city, state and ZIP code. If the envelope is not business stationery, devote two lines in the upper left to your street address, and city, state and ZIP code.

# 5D Good-News Correspondence

## Purpose, Audience and Media

A good-news message concisely conveys information that will please the recipient.

The audience of a good-news message usually is one person. Occasionally, the good news may apply to an organization, such as an organization that has won an award. In that case, the recipient is a representative of that organization.

Good-news messages generally are sent via business letter or email. Business acquaintances do use email, or even a phone call, for minor good news. However, email lacks the formality and keepsake value of a traditional letter on high-quality stationery. If the good news is important, or if you're sending it to someone you don't know well, be traditional and use paper (unless you know that the recipient prefers email or a phone call).

**KEY TO SUCCESS**

A good-news message should announce the good news in its first paragraph.

## Format/Design

Follow the general guidelines for the business-message format on pages 297–301.

## Content and Organization

The good-news message has three or four parts that usually translate into three to five paragraphs.

### *Part 1 of 4: Deliver the Good News*

Open positively. Deliver the good news immediately. Or, if the recipient first wrote you with a request, you can thank them for contacting you. Then announce the good news: for example, a refund. If the good news involves an award, a promotion or something similar, offer congratulations. If you open with congratulations, be sure to immediately inform the recipient about the honor or award. Otherwise, your message may sound as if you're congratulating them for something that you assume they already know.

If your good-news message is in response to a complaint, begin, as noted earlier, by thanking the individual for contacting you. Then, in one sentence, be understanding but don't apologize in a way that accepts blame. For example, you can write, "I regret that you're dissatisfied with the quality of our service." After the apology, concisely announce the good news.

Everything discussed in these "deliver the good news" guidelines could be included in a three- or four-sentence opening paragraph. Do not include specific details about the good news, and do not make requests (such as asking the recipient if they will accept an award). The only function of this first paragraph is to announce the good news and, if appropriate, to offer congratulations.

### *Part 2 of 4: Explain the Details*

In a new paragraph, explain the details of the good news. For example, will you issue a refund check? If so, when will it be sent? Are you inviting the recipient to an awards banquet? If so, when and where will it be? In this section, inform the recipient of any details they should know to take advantage of the good news. If this information includes a request to contact you, you may wish to save that until the last paragraph of the message. This section can include more than one paragraph if needed.

### *Part 3 of 4: Say What the Good News Means to You (Optional)*

If appropriate, in a new paragraph you may discuss your feelings (and/or the feelings of your organization) about the good news. For example, if your message offers the recipient a job, this paragraph could mention how delighted you are to offer this job and how you look forward to working together.

### *Part 4 of 4: End Positively, Perhaps With Instructions*

In your final paragraph, be courteous and positive. Consider specifying what the next action should be. Include details about how the recipient can contact you if necessary. If you are addressing a situation brought to your attention by the recipient, you may thank them again for contacting you. If congratulations are appropriate, you may again express them. Include a standard "Sincerely" sign-off.

An email version of the good-news example on page 305 would begin with the words "Dear Patty." A subject line might be "Congratulations!" or "2022 Supplier of the Year!"

## GOOD-NEWS TIPS

1. **Focus on the recipient**: Don't write at length about you and your organization unless the recipient needs or desires that information.
2. **Be understanding**: If you're responding to a negative situation with good news, empathize. How would you feel if the roles were reversed?
3. **Protect your organization**: When responding to a negative situation, never disparage your organization.
4. **Don't overpromise**: When responding to a complaint, know what you can offer and what you can't. If you don't know, ask the appropriate person in your organization. Make no commitment on behalf of your organization that it can't or won't keep.
5. **Be polite**: Remember that you represent your organization. Be courteous—even if you haven't been treated courteously by the recipient.

## Good-News Letter

2010 Ridglea Dr. • Austin, TX 55111 • 555-999-5555

January 18, 2023

Ms. Patricia Morris
Chief Executive Officer
Texas Mix Productions Inc.
1989 Binkley Ave.
Dallas, TX 87538

Dear Patty,

Congratulations! It is my great pleasure to tell you that MGS Interactive Games has named you and Texas Mix Productions its 2022 Supplier of the Year. The competition for this honor was impressive, but at its January meeting, our Management Council reviewed the list of finalists and selected Texas Mix Productions as our top supplier.

MGS Interactive Games invites you to be a guest of honor at our annual awards banquet March 15 at 7 p.m. at the Grand Royale Hotel in Austin. We'll gather for cocktails at 6 p.m. in the Casino Room. You are more than welcome to bring up to four guests. (If you'd like to bring a few more, we can certainly arrange that.) Please contact me to clarify what arrangements would be best for you.

At the banquet, David Mertz, our vice president of acquisitions, will present our Supplier of the Year 2022 trophy to you. We'll then invite you to speak for two or three minutes. Your speech needn't be long or elaborate; we just want another chance to cheer for you.

I hope you know what an honor it is for me to write this letter. When we were roommates 10 years ago at Palmquist University, who knew our paths would cross again at such a wonderful moment? I'm proud to claim you as a business associate and a friend.

Again, congratulations on this significant achievement in your career. Please phone me (555-999-5618) at your earliest convenience to confirm a plan.

Sincerely,

*Janet*

Janet Walker
Special Events Coordinator

# 5E Bad-News Correspondence

## Purpose, Audience and Media

A bad-news message tells the recipient something that they doesn't want to hear: for example, no refund or no job openings at this time.

The audience of a bad-news message usually is one person. Your organization often wants to maintain a good relationship with that person—or, at least, your organization wants to keep that person's goodwill. The need to preserve a good relationship while delivering bad news can make the writing of a bad-news message difficult.

Unlike good-news messages, bad-news messages rarely have keepsake value. Bad-news letters still exist, but such messages increasingly use email for delivery. (However, don't use email to avoid a face-to-face bad-news meeting when such directness, however embarrassing, would be the most ethical way to deliver the bad news. A bad-news email could follow such a meeting to ensure a record of the details.)

**KEY TO SUCCESS**

A well-organized bad-news message explains the reason(s) for the bad news before it announces the bad news.

## Format/Design

Follow the general guidelines for the business-message format on pages 297–301.

## Content and Organization

The bad-news message has five parts that usually translate into three paragraphs.

## *Part 1 of 5: Begin Courteously, Focusing on a Positive Relationship*

Part 1 usually is one paragraph. Thank the recipient for contacting you, if that's appropriate. Because you want to maintain a relationship with the recipient (or at least keep their goodwill), express your appreciation for that relationship and/or your respect for the recipient. For example, to a customer you might describe how much you appreciate the opportunity to serve them. To a job applicant, you might note how flattered your organization is to receive applications from high-quality candidates.

*In this opening paragraph, do not mention the bad news.* Do your best to focus on the value of the relationship. Don't talk about yourself or your organization too much. By discussing the recipient's role in the relationship, you show that you value them. In most bad-news situations, three or four sentences suffice to create this polite beginning.

## *Part 2 of 5: Explain the Reason(s) for the Bad News*

Part 2 usually is the beginning of the second paragraph. It can be more than one sentence. This concise description of the reason(s) for the bad news is one of the most strategic parts of the bad-news message: You offer an explanation for the bad news *before* you deliver the bad news. Therefore, if the bad news angers the recipient, at least they understand your reasoning. The recipient may even understand your reasoning so well that they graciously accept the justice of the bad news.

For example, before telling a job applicant that you can't hire them, first explain that you have no available positions for someone with their qualifications. Deliver the explanation before you deliver the bad news.

The explanation of the bad news comes at the beginning of the second paragraph and is an important transition in the message. The explanation is a bridge between the relationship-focused first paragraph and the upcoming delivery of the bad news. This explanation should avoid a sudden shift to a harsh tone from the courtesy of the first paragraph. An abrupt, stern tone at this point would make the first paragraph seem insincere.

## *Part 3 of 5: Deliver the Bad News*

In the same paragraph as the explanation (usually the second paragraph), state the bad news clearly and concisely—in one sentence, whenever possible.

Don't set off the bad-news sentence as its own paragraph. One-sentence paragraphs get extra emphasis, and you don't want to emphasize the bad news. Ideally, the bad-news sentence will appear in the middle of the second paragraph, and middles are points of low emphasis. The bad news won't get the extra emphasis that paragraph openings and closings get.

## *Part 4 of 5: Cap the Bad News With Something Neutral or Positive*

As noted previously, don't let the bad-news sentence close the paragraph. Don't let it gain extra emphasis by echoing into the momentary silence that follows the end of a paragraph. Instead, "cap" the bad news. That is, close the paragraph with something neutral or positive.

For example, if you're turning down a job applicant, follow the bad-news sentence with the promise that you'll keep their résumé on file for one year and will be in touch if any suitable jobs open.

Again, note that parts 2, 3 and 4 often go in the same paragraph: first the explanation; then the bad news; then the neutral or positive cap. This organization places the bad news in the middle of the paragraph, a point of low emphasis.

## *Part 5 of 5: Close Courteously, Focusing on a Positive Relationship*

Sounds a lot like Part 1, doesn't it? It should: Just as you began the bad-news message by focusing on a good relationship between your organization and the recipient, close the message in the same way. Show the recipient that they have value to your organization.

This closing is usually one short paragraph, the last paragraph of the bad-news message. *Do not refer to the bad news in this paragraph, not even indirectly*. You delivered the bad news clearly in the previous paragraph. Don't emphasize the bad news by repeating it, or even referring to it, in this closing paragraph. Instead, discuss the positive aspects of the relationship or a continuation of the relationship, or focus on preserving the goodwill of the recipient. Let the recipient finish reading the message with a vision of a good relationship. Include a standard "Sincerely" sign-off.

## *Exceptions to the Standard Bad-News Message Organization*

The five-part bad-news message, described previously, works in almost every bad-news situation. However, when the bad news is devastating and recipient(s) already know some of the details, you might want to consider a different organizational strategy. This different strategy quickly and boldly confronts the bad news in order to show that your organization takes the situation seriously. For example, a CEO who must explain plummeting stock prices might want to use this more direct strategy in a message to stockholders or in the company's annual report:

- In the first paragraph, immediately announce the bad news. A brief explanation can come before the bad news, but you must announce the bad news in the first paragraph.
- In a new paragraph or paragraphs, discuss in detail how the situation happened and what your organization is doing to improve the situation. Show that you understand and are addressing the problem.

- In the closing paragraph, express confidence about the future and announce any communication actions that should follow. Should the recipient(s) contact you? Will you send periodic e-blasts to the recipients? What other communications actions do you recommend to help address and resolve the bad news?

An email version of the bad-news example on page 310 would begin with the words "Dear Ms. Jones." A subject line might be "Thank you for proposal" or "Relationship with Brand Z."

## BAD-NEWS TIPS

1. **Don't point fingers**: As much as possible, avoid personal pronouns (any form of *I* and *you*) in the explanation for the bad news and in the bad-news sentence—for example, "*We* are not renewing *our* contract with *your* company." Those pronouns encourage the recipient to take the bad news personally, and you don't want them to do that. You want to preserve the relationship or at least the recipient's goodwill.
2. **Don't be wordy**: Don't attempt to hide the bad news in an avalanche of words. Avoid being blunt to the point of rudeness, but be concise.
3. **Don't vent**: Don't let personal feelings, such as anger or sympathy, excessively influence your message. Your job is to protect and promote your organization by effectively managing the relationship with the bad-news message's recipient. You might hurt that relationship by being overly emotional. For example, if you sympathize too much, you might sound as if you disagree with your organization's actions.
4. **Never disparage your own organization**: You may be tempted to defuse a recipient's anger by agreeing with them, but avoid writing something like "I agree that we didn't perform very well."
5. **Be careful with apologies**: You can apologize to a recipient, but avoid doing so in a way that accepts legal responsibility or blame (unless your organization's legal team gives you permission to do so). Avoid writing, "I apologize for our poor performance in this area." Instead, you might write, "I regret that you're unsatisfied with the quality of our service."
6. **Avoid clichés**: Don't write "It has come to my attention," "I regret to inform you," "Pursuant to your request" and similar sentences. These worn-out passages are so overused that they sound insincere.
7. **Use for other documents**: Consider using the five-part bad-news organizational strategy in appropriate memos and speeches.

## Bad-News Letter

2010 Ridglea Dr. • Austin, TX 55111 • 555-999-5555

November 2, 2023

Ms. Molly Jones
President
Brand Z Software Systems
1989 Lancaster St.
Dallas, TX 76450

Dear Ms. Jones:

Thank you for your recent proposal explaining how Brand Z Software Systems could contribute to the new production processes at MGS Interactive Games. We thought your proposal was innovative, and it showed us why Brand Z is such a respected company in the software industry.

When MGS Interactive Games launched the new production processes, we signed a long-term contract with NW5 Software Supply Systems to maintain and monitor those processes. Therefore, MGS cannot offer Brand Z a maintenance contract at this time. However, MGS is continually expanding our lineup of interactive games, and we would be pleased to consider any proposals from Brand Z regarding future production processes.

Again, thank you for your interest in MGS Interactive Games. I would welcome an opportunity to meet with you and discuss future business opportunities.

Sincerely,

*Mark Smith*

Mark Smith
Executive Vice President

# 5F Job-Request Correspondence

## Purpose, Audience and Media

A written job request is a one-page business letter or a concise email in which the writer asks the recipient for a job.

The audience of a job-request message usually is one person—a busy person who may be receiving many similar requests. This recipient may be nervous about the upcoming hiring decision. Although the job search can be nerve-wracking for you (the request writer), the search also can be stressful for the person who screens candidates and conducts job interviews. If the interviewer makes a good decision in hiring you, that person has improved their stature within their organization. The organization's leaders see that the interviewer can recruit talented individuals. However, if the interviewer hires someone who doesn't perform well, the leaders may question that person's judgment and abilities as a manager. Therefore, the recipient of job-request correspondence wants something almost magical: They want a message that seems to promise that the writer would be the perfect employee.

The purpose of this section is to help you present yourself as that employee.

Again, a job-request message can be sent via email, an attached file, a website application form or a sheet of high-quality paper that you'll put in an envelope and place in the mail. Employers often specify their preferred format.

If you know that an employer prefers an email request, you can skip the headings of a business letter and begin with the salutation (*Dear Ms. Jones:*). After that salutation, follow the four-paragraph method described below.

If you send a formal letter as an attachment to an email message (or, more common, if you attach your résumé), be certain to scan that document for viruses. Emailing an infected document to a potential employer is a sure way to get fired before you're hired.

### KEY TO SUCCESS

In a successful job-request message, you should include specific research to explain why you want to work for that particular organization. Show the recipient that this isn't a form message, one you've sent to a dozen other potential employers. With your research, show the prospective employer that you seek more than just a job; show them that you truly know their organization and want to work for it.

## Format/Design

For paper letters, follow the general guidelines for the business-letter format on pages 297–301. However, unless you design your own stationery, your job-request letter will be on a blank sheet of paper. Therefore, you need to include your return address. Include that address at the top. Instead of beginning with the date, begin with your street address, then city, state and ZIP code. The first three lines of your letter, single-spaced and placed in the upper-left corner, would look like this:

712 Custer St.

Paderno, TX 80476

January 18, 2023

Note that in this new heading, you do not include your name. (You will type and sign your name at the bottom.) After this heading, skip one line and then follow the traditional headings for a business letter, beginning with the recipient's name.

At the bottom of the letter, do not include a title under your typed name.

## Content and Organization

The job-request message has four parts that usually translate into four paragraphs. In each paragraph, avoid sounding like a business-writing machine. Strive for a tone that shows you're an intelligent, articulate, friendly, ambitious job seeker.

### *Part 1 of 4: Tell the Recipient Why You're Writing*

Begin by saying why you're writing and who you are. For example, you might begin this way:

> Please accept this as my application for a position as account assistant at Jones & Jones. This May I will receive a bachelor of science degree, with honors in journalism, from Palmquist University. I share the commitment expressed in the Jones & Jones mission statement: "Providing services that exceed expectations." That's why I would like to join the team at Jones & Jones.

### *Part 2 of 4: Explain Why You Want to Work for This Particular Organization*

In this paragraph, demonstrate that this isn't just a standard message that you've sent to a dozen potential employers. Use specific knowledge of the organization to explain your eagerness to work there. To gain this knowledge, you'll need to do more than just examine the organization's website. Search online news databases for information about the company, such as awards and new projects. Learn where the company has been, where it is now and where it hopes to go. Your second paragraph might sound something like this:

> My interest in working for Jones & Jones began with your handling of the Fat Burger account. You not only won a Bronze Quill Award for excellence but also helped set record profits for that restaurant. Your pro bono work for City Children's Center has inspired our entire community, including me. And your recent remarks about business ethics at the Chamber of Commerce luncheon show why clients and competitors alike respect Jones & Jones for integrity as well as excellence. I'd like an opportunity to contribute to the continuing success of Jones & Jones.

Chances are, this paragraph will grab the recipient's attention because that individual is used to receiving standard messages that could have gone to any organization. This second paragraph will help your job request stand out for a variety of reasons:

- You didn't write a so-called form letter or email.
- In gathering organization research, you worked harder than other job applicants.
- You used a smarter, more sincere approach than other job applicants.
- You showed that you want more than just any job; you really want to work for the recipient's organization.
- You flattered the recipient by showing specific, well-informed interest in their organization.
- You eased the recipient's concerns about the hiring process. You've shown yourself to be a smart, hard-working and diplomatic candidate who will enhance the recipient's reputation for recruiting good employees.

## *Part 3 of 4: Describe Your Specific Accomplishments*

In the third paragraph, sell yourself. Be specific about your accomplishments and what you can bring to the organization. Name former employers and particular successes. When possible, focus on results. Creating a content marketing campaign is impressive, but creating a content marketing campaign that doubled the number of an organization's social media followers is better. It's fine to repeat parts of your résumé here:

> I believe that I have the skills to be part of the Jones & Jones success story. As I hope you'll see on my résumé, I. . . .

### *Part 4 of 4: Show Initiative: Ask for an Interview*

In the fourth and final paragraph, close by showing initiative:

> I will call next week to see if we can schedule an interview. Thank you very much for your time and consideration.
>
> Sincerely,
>
> *[Be sure to sign the letter legibly.]*
>
> Your typed name

By asking for the interview, you show polite aggressiveness, a good quality in a prospective employee. You don't simply say, *I hope to hear from you at your earliest convenience*. Instead, you show that you're the kind of person who tries to make good things happen. Rather than just hoping for an interview, you're willing to step up and ask for one. That polite aggressiveness might appeal to the recipient, who will think you'll carry that same initiative into the workplace. Include a standard "Sincerely" sign-off.

If a follow-up call is impossible or inadvisable, consider concluding with "Thank you very much for your time and consideration. I hope we'll have a chance soon to discuss opportunities at Jones & Jones."

## JOB-REQUEST TIPS

1. **Proofread**: And then proofread again—and have others proofread the message. Just as with the résumé, a single error can be fatal.
2. **Make the call**: Have a set speech ready when you make your follow-up call to try to schedule an interview. Remind the recipient of your letter or email, and ask if an interview is possible. If you get voicemail, be ready with your concise, professional speech. One more call-back is all right, but don't leave more than two messages. If you get an assistant, explain why you're phoning and ask for the recipient of your job request to return your call. Be polite. An assistant's opinion of you can be very influential in whether you'll be hired.
3. **Or don't make the call**: Don't telephone an organization that has specified "no telephone calls" in its employment advertisements. In such cases, as noted before, the last paragraph of your message might be simply "Thank you very much for your time and consideration."
4. **Prepare for your interview**: A good approach is to have a two-topic agenda for the interview: to show that you know the company and to show that you have the skills to do the job. When possible, steer your answers to those two areas. Do even more research on the organization, and prepare answers for potential questions. For example, what are your strengths? Your weaknesses? What's the most

## JOB-REQUEST TIPS

rewarding thing you've ever done? What are your pet peeves? Where do you want to be in 10 years? Be ready with questions of your own—questions that display your knowledge of the organization. Don't ask about salary and benefits; let the interviewer introduce those topics. An excellent resource for job interviews is the book *Knock 'Em Dead: The Ultimate Job Search Guide* by Martin Yate.

5. **Prepare a portfolio**: Have a well-organized, diverse, professional-looking collection—both online and tangible—of your work. Online portfolios generally have a table of contents on the homepage. The first page of a tangible portfolio is usually your résumé, with the second page being a table of contents (without page numbers) that names each section. In both formats (online and tangible), organize your portfolio by document categories. In a tangible portfolio, you can use a divider tab for each section—for example, news releases, newsletter stories, video advertisement scripts and so on. Tangible portfolios often are three-ring binders with zippers that help seal the notebook to prevent media kit folders or bound proposals from tumbling out. Use transparent sheet protectors to enclose your documents.
6. **Dress for success**: Dress appropriately for the interview. Select clothing in which you look and feel professional and comfortable. For job interviews, it's better to be overdressed than underdressed.
7. **Project confidence**: During the interview, have a firm handshake and maintain eye contact. Don't fidget. Fold your hands in your lap if necessary.
8. **Be thankful**: After the interview (on the same day), send a brief thank-you email in which you (1) thank the recipient for the interview, (2) mention something specific that you appreciated learning during the interview and (3) gracefully ask for the job. Write to everyone who interviewed you. Vary your wording so that the messages are not duplicates. Also write to anyone who helped you set up the interview, including secretaries and friends who recommended you. (Some job-advice sources still recommend sending handwritten thank-you notes. Such notes are impressive, but unless you can ensure that they arrive within 12 hours of the interview—a difficult feat—you're well-advised to use email.)

## Job-Request Letter

712 Custer St.
Paderno, TX 80476
January 18, 2023

Ms. Ivy Jones
President
Jones & Jones
1876 Lancaster St.
Dallas, TX 76450

Dear Ms. Jones:

Please consider this letter to be my application for a position as account assistant at Jones & Jones. This May I will receive a bachelor of sciences degree, with honors in journalism, from Palmquist University. I share the commitment expressed in the Jones & Jones mission statement: "Providing services that exceed expectations." That's why I would like to join the team at Jones & Jones.

My interest in working for Jones & Jones began with your handling of the Fat Burger account. You not only won a Bronze Quill Award for excellence but also helped set record profits for that restaurant. Your pro bono work for City Children's Center has inspired our entire community, including me. And your recent remarks about business ethics at the Chamber of Commerce luncheon show why clients and competitors alike respect Jones & Jones for integrity as well as excellence. I'd like an opportunity to contribute to the continuing success of Jones & Jones.

I believe that I have the skills to be part of the Jones & Jones success story. As I hope you'll see on my résumé, I wrote a multimedia news release for the Flora County United Way that generated coverage from seven media outlets in the county. I've produced eight video oral histories for Central City Senior Center and have designed and written more than two dozen digital ads for the Palmquist University Daily News. During my internship at United Marketing Corp., I helped prepare seven direct-marketing packages, and I helped write and edit six video news releases, all of which were used in regional newscasts. I'm a multimedia writer who gets results, and I'd like a chance to prove that at Jones & Jones.

I will call next week to see if we can schedule an interview. Thank you very much for your time and consideration.

Sincerely,

*Taylor Jackson*

Taylor Jackson

# 5G Résumés

## Purpose, Audience and Media

A résumé is a short document that summarizes an individual's education, professional experience, professional abilities and other work experience. A résumé can include information on individual honors and on activities. Generally, a résumé is one page.

A job seeker sends a résumé to a potential employer. The job seeker also typically sends a job-request email or letter (also called a résumé cover letter; see pages 311–316), a list of references and some work samples.

A traditional résumé exists as a single sheet of paper. However, résumés can go online in three different ways. You can send an email message requesting a job and include your résumé as an attached document. Use this method only if you know that the employer accepts online job requests. If you include your résumé as an attached file, include it as a PDF whenever possible. Also, be sure to scan the file for viruses. An infected file either will be destroyed by the employer's safeguards—meaning no résumé to read—or it will infect your potential boss's computer. Either way, you lose.

Résumés also can go online through job-search websites, such as Monster.com. Each site has its own instructions for how to create and place your online résumé.

Finally, if you have your own website, consider posting your résumé—plus impressive items from your portfolio—there.

**KEY TO SUCCESS**

A successful résumé is well-organized, specific and concise.

## Format/Design

The design or layout of a résumé is often a prospective employer's first impression. A résumé's appearance and organization immediately send a message about how well you communicate. Experts don't agree on one perfect format for a résumé. As many students know, if you ask three professors and three professionals how to organize a résumé, you may get six different answers. In this chapter, we'll recommend a traditional format that works well, one that allows you to pack in a great deal of information—but attractive résumé formats abound online. Select one that pleases you and offers an easy-to-follow presentation to potential employers.

The easiest way to understand this traditional format is simply to examine the sample résumé on page 321. Try to keep one-inch margins at top, bottom and both sides. To squeeze in more information, you can cut those to three-quarters of an inch. Certainly don't go lower than a half inch. A common typeface and size is 11-point Times or Times New Roman. Set the line spacing to 11 points or even 10 points. If you still lack space, consider an even smaller size of a sans-serif typeface such as Geneva. Let your eye be the judge. A résumé can be tight with information, but it shouldn't look overloaded. Again, keep your résumé to one page.

Single-space your résumé, and insert an extra space before each new section. Use boldface type to highlight key words, such as categories of information (for example, **Education**) and names of organizations that employed you. However, don't overuse boldfacing. If everything seems to be highlighted, then nothing gains emphasis.

Your résumé can include the following categories of information, often in this order: Education, Professional Experience, Other Employment, Skills, Honors and Activities. In the sample on page 321, note how those titles can appear boldface on the left side. This technique clarifies the organization of your résumé.

For traditional paper résumés, use a conservative color (white, gray or cream) unless you know that an employer seeks flamboyance and wild creativity. Use good cotton-fiber paper, not just photocopy paper.

## Content and Organization

Put your name and contact information (address, phone number and email) at the top. Students occasionally include both a school address and a permanent (family) address.

Beneath the contact information, summarize your "Education." Don't include information about high school. List all universities at which you've gained college credit, including any universities you attended during study abroad experiences. Include dates for each university. Include your graduation date or anticipated graduation date. Potential employers generally expect to see your grade-point average, but it's not required. Chances are, employers will be more interested in your experience and your portfolio of professional work than in your GPA. If you list your GPA, show it as a ratio: 3.4/4.0.

Next comes "Professional Experience." In this category, include jobs since high school graduation that are relevant to your career goal. Be sure to include professional internships here. List them in reverse chronological order, beginning with the most recent. For each job, list your title and describe your duties. For each duty, begin with a strong, specific verb—or verbs. For example, don't write, "Helped prepare social media calendar" or "Responsible for social media calendar." Instead, write, "Wrote and scheduled posts for social media calendar." If you still are employed in the job, use present-tense verbs; otherwise, use past-tense verbs. For each job, list your dates of employment. Listing those dates in the left margin allows an employer to easily scan the dates of your employment history to see if there are any long gaps.

In "Other Employment," include other jobs since high school graduation. These jobs may not relate directly to your career goal, but they help show work ethic, versatility and the ability to get along with others. This may include military service. Employers know that you're a college student. They'll expect to see jobs such as waiter, store clerk and lifeguard. List these jobs in reverse chronological order, beginning with the most recent. For each job, list your title and describe your duties. Again, use strong, specific verbs to describe your duties. Be sure to include any duties that show that an employer trusted you with money or with the business itself. For example, if you opened or closed the business, or if you totaled cash registers or trained other employees, be sure to include that information. For each job, list your dates of employment, just as you did in the previous section.

If you include a "Skills" section, list only skills that separate you from the normal employee. For example, don't include "Proficient in Microsoft Word." That's a professional expectation, not a skill. Instead, include such skills as "Proficient in Photoshop, InDesign and Excel." Include fluency in foreign languages in this section.

If you include an "Honors" section, be concise. List only the name of the award and the date. If an explanation is necessary, consider including just the name of the organization that bestowed the honor.

Like an "Honors" section, an "Activities" section is not required. Include it only if space allows and if you believe the rest of the résumé fails to show that you're a well-rounded person. In "Activities," you can include organizations to which you belong and sports or other activities to which you devote time. As always in a résumé, be concise. Include just the names of organizations and activities.

## RÉSUMÉ TIPS

1. **Be concise**: Keep a paper résumé to one page. More than one middle-aged employer has said, "If I can get my résumé onto one page, don't tell me that a college student can't do the same."
2. **Proofread**: Proofread. Proofread. Have others proofread your résumé. Then proofread it again. Proofread your résumé backward, one passage at a time. A single typo can be fatal to your job search.
3. **Don't automatically include an objective**: If a job-request email or letter accompanies your résumé, you probably don't need a "Career Objective" just below the contact information at the top. That job-request message will specify your career objective. However, if a job-request message *doesn't* accompany your résumé, then a "Career Objective" can be a good idea. Put that section just below the contact information at the top. Treat the words "Career Objective" in the same way you treat other labels, such as "Education." In your objective, be concise. Don't be pretentious, and don't be self-centered, focusing only on what you hope to gain. A good career objective comes right to the point: "An entry-level position in events management."
4. **Consider paragraphing**: Consider saving space by listing job duties, awards, honors and activities in paragraph form rather than giving one line to each entry. If you have space for one line for each of those items, consider introducing each item with a small bullet (•). However, as you gain more experience, chances are good that you'll lack the space for one line per item.
5. **List references**: If you have room, consider listing references on your résumé. However, you may lack room. It's generally all right to list your references in a separate document labeled "References." Under that label, put your name and contact information, just as you did on the résumé. For each reference, include name; title; organization; street address; city, state and ZIP code; phone number, including area code; and email address. If you're working with paper, don't staple this sheet to your résumé; don't create the appearance of a two-page résumé. Some employers provide specific instructions regarding references.
6. **Don't list high school achievements**: As noted before, don't include any information from your high school years. Some employers have the perception, fair or not, that high school students are not adults. Employers are interested in your achievements as an adult.

# Résumé

**John Doe**

**Current Address**
3309 Frontier Rd.
Tonsing, FL 98638
(555) 253-9876
jdoe@ucfla.edu

**Permanent Address**
4974 47th St.
Ink City, OK
(555) 873-6524

**Education**

University of Coastal Florida, Tonsing, Florida
B.S. in Journalism, with emphasis in Strategic Communication
Degree expected: May 2024
Current Grade Point Average: 3.85/4.0
City University, Paris, France
August–December 2022

**Professional Experience**

Jan. 2023–present
**Advertising Intern**, Carter and Associates, Midway, FL
Develop strategic message planners for print, mobile and web ads. Write copy for print, mobile and web ads. Assist with research and writing of proposals for clients. Prepare slides for proposal presentations.

Jan. 2022–present
**Freelance Projects**
Write newsletter stories for Compton County Senior Services newsletter. Write multimedia news releases for Tonsing City Volunteer Fire Department. Write and produce podcasts for Midway Adult Learning Center.

Jan. 2022–Aug. 2022
**Corporate Communications Intern,** Coastal Power, Tonsing, FL
Wrote and distributed multimedia news releases. Wrote and edited weekly online newsletter. Wrote status updates for company Facebook page.

**Other Employment**

Sept. 2021–Jan. 2022
**Store Clerk and Cashier,** Fashion Fun Stop, Tonsing, FL
Opened store. Closed store. Ordered inventory. Totaled cash drawers. Assisted approximately 30 customers per day. Named Employee of the Month two times.

May 2021–Aug. 2021
**Waiter,** Old Doug's Fishing Shack, Ink City, Okla.
Assisted approximately 40 customers per day. Trained new employees. Edited new menu.

**Skills**
Proficient in Photoshop, InDesign, Excel and digital video-editing systems. Fluent in French.

**Honors**
William Allen Bowen Academic Scholarship (three years). School of Journalism Dean's List, 2022, 2021, 2020.

**Activities**
Public Relations Student Society of America. University of Coastal Florida Ad Club. St. John's Catholic Campus Center volunteer. University of Coastal Florida Water Skiing Team.

# 5H Memoranda

## Purpose, Audience and Media

A memorandum is a written message to an internal audience—that is, to a person or people within your organization. Think of a memo as an in-house letter. Memos are usually informal and conversational, though not needlessly wordy. The memorandum can incorporate other forms of business writing. A memo can be a news announcement (pages 103–106) or a modified good-news or bad-news message (pages 302–305; 306–310).

The audience for a memo, again, is a person or people within your organization. You can send memos to people outside your organization if you have a longstanding business relationship with them and you know them well. Memos are less formal than business letters and even business phone calls. Send memos outside your organization only when your relationship with the recipient allows such informality.

Memos generally are sent via email. However, some memos still are written on paper and distributed through office mail. Paper memos sometimes serve as a cover sheet for other paper documents, such as a business report.

**KEY TO SUCCESS**

A good memo shows respect for the reader. It announces its message in the subject line and comes to the point quickly and gracefully.

## Format/Design

The headings of a paper memo should specify the recipient, the sender, the date and the subject. Email, therefore, is an effective medium for memos. Email systems automatically specify the sender's name and address as well as the date. In addition, they prompt the sender to specify the recipient and the subject.

Some organizations still use paper memos when a message is important, will be frequently consulted or must be filed for legal records. Such organizations may still have a memo form that reminds you to include necessary information about:

- The date
- The recipient (a *To* section)
- The sender (a *From* section)
- The subject

At the top of a paper memo, after you record the date, record the recipient's name and title in the *To* section.

In the *From* section, record your name, your title, your office location, your department name, your phone number and your email address. Including all this *From* information is important. It provides the recipient several ways to contact you. All the *From* information even allows the recipient to write a brief answer on your memo, cross out the original *To* and indicate that the memo should be returned to you at the address listed in the *From* section.

If you send a paper memorandum to more than one person, include all the appropriate names and titles in the *To* section. You can, however, "cc" a memo if you want someone to see a memo that you've sent to someone else (see the business-letter guidelines, page 298). You also can simply write "All Employees" in the *To* section.

Always include a brief description of the subject in the memo's heading; this is true for email as well as paper memos. (If you're writing a bad-news memo, strive for a neutral but accurate subject description.) The subject description usually is not a complete sentence.

Standard headings for a paper memo are single-spaced with an extra space between each section. Headings for a paper memo can look like this:

**Date:** Nov. 20, 2023

**To:** Melva Young, Director of Personnel

**From:** Mike Smith<br>Director of Communications<br>Communications Dept.<br>Building H, Office 427<br>Ext. 3875<br>msmith@mgsintgames.com

**Subject:** Employee Handbook revisions

Single-space the paragraphs of a paper memorandum. Double-space between paragraphs. Don't indent paragraphs.

After you have carefully proofread a paper memo, write your initials next to your name at the top of the page. This is how you sign a paper memo.

## Content and Organization

In email memos, the most common salutation is the recipient's name plus a colon. For example, "Sarah:" would be the salutation to a co-worker you know well. (If you write "Dear Sarah," you would use a comma.) Salutations such as "Ms. Hernandez:" or "Mr. Fulton:" are suitable for more formal situations.

The first sentence of a memo generally moves quickly to the subject of the memo.

Because memos are less formal than business letters, they usually lack the "Sincerely" sign-off. A simple "Thanks!" or "Thank you" often suffices for a closing paragraph.

Below the "Thank you" in an email, simply type your name. If your salutation was simply the recipient's first name, then type only your first name. However, if your salutation was more formal, such as "Ms. Hernandez:" then type your first and last names.

Ideally, you have used the options of your email program to automatically add your signature (name, title, address and phone number) to the bottom of your email messages. This information would appear below where you have typed your name. (Because of the automatic information, your name may appear twice at the bottom of an email message. No problem—that's traditional.)

Paper memos are different from email memos. They have no salutation; the *To* information at the top of the memo serves as the salutation. Therefore, after completing the headings, begin the memo with the first sentence of your message.

Paper memos also have no sign-off—no "Sincerely" line—nor do they have a signature and/or typed name at the bottom. Instead, the *From* information at the top of the memo, which includes your handwritten initials, serves as your sign-off and signature. Because of this lack of a traditional sign-off, the last words of your final paragraph should provide a sense of closure. Thanking the recipient or suggesting what the next communications action should be, or both, can provide closure.

### *Position Memos*

Position memos are prepared for an organization's leaders. Position memos address issues that may require action. This type of memo often has a concise, three-part organization:

- Paragraph one: What is the issue? What is the problem? This section often is labeled "The Issue."
- Paragraph two: What are the possible solutions, courses of action or policies? This section often is labeled "Possible Actions."

- Paragraph three: What solution, course of action or policy do we recommend—and why? This section often is labeled "Recommendations" or "Recommended Position."

Position memos are concise. Consolidating so much information into three paragraphs demands skill in precision and editing.

## *Career Advancement Memos*

In the following workplace situations, you can use memos to advance and protect your career:

- **Weekly updates**: If your organization doesn't have a standard project-status meeting or form, get in the habit of writing a memo to your boss every Friday before you leave for the weekend. In that memo, briefly describe the status of all your projects. If you foresee problems with any of your projects, use your weekly memo to say so. If you announce a problem, also try to propose one or two solutions. Such memos often are called progress reports. Keep copies of these memos. If a problem does arise, you will at least have a record of your warnings.
- **Project updates**: Similar periodic memos can be sent to clients, updating them on the progress of projects. Such memos can be particularly useful if you have a client who misses deadlines or whose lack of performance is affecting the quality of a project. Never threaten a client in such a memo.
- Diplomatically point out the consequences—financial and otherwise—of their actions. Such memos leave a record, showing that you've done your best to act in your organization's interest. You may wish to consult your boss before writing such memos.
- **Memos of concern**: If a client or a boss asks you to act unwisely or, worse, unethically, and you're concerned that failing to comply will hurt your career, write that client or boss a memo detailing the requested action and ask if you understand correctly. If circumstances allow, consult your boss before writing such a memo to a troublesome client. Seeing the request in writing also may help the client or boss perceive the unwise or unethical nature of the request. Or, better, perhaps you have misunderstood the situation, and the memo will help resolve the confusion.
- Such a memo increases the chances that the matter will be dropped—because, again, the memo begins to create a record of the situation. A business letter can fulfill the same purpose with an external client. If the client or boss responds to your memo with a phone call or personal visit, be sure to write yourself a memo about that conversation (see the next paragraph).
- **Work-related memos**: If you're in an unpleasant situation at work, such as giving an employee a poor evaluation—or, worse, having to fire an employee—write a memo to yourself, simply for your own files, describing what happened. Be brief but accurate and detailed. If you're writing a paper memo, be certain to date it. That memo can

help refresh your memory later if you need to describe the situation to internal or even external authorities.

## MEMO TIPS

1. **Be diplomatic**: Although memoranda usually are internal documents, don't write anything in a memo that you wouldn't want to see as the top headline in the *New York Times* or as the lead story on CNN. Internal information has a way of becoming external. Be careful how you phrase bad news and sensitive topics. An effective test is to ask yourself, "What damage could my worst enemy or my organization's worst enemy do with this memo?" (This is good advice for writing any sensitive document, not just a memo.)
2. **Get to the point**: Effective memos are concise and precise.

## Paper Memorandum

2010 Ridglea Drive • Austin, TX 55111 • 555-999-5555

**MEMO**

Date: Oct. 15, 2023

To: All Employees

From: Clarence Emsworth CE
Director of Information Technology
IT Dept.
Blandings Building, Office 427
Ext. 3875
ce@mgsintgames.com

Subject: Passwords to proprietary software

Attached to this memo is a wallet-sized card with temporary passwords for seven of the proprietary programs MGS uses in game design.

As you know, ordinarily these passwords are encrypted and stored online and can be accessed with your employee password and our traditional security protocols.

As we upgrade network security, however, we have temporarily removed those passwords–thus this card. We ask that you not share this card with others outside our company. We anticipate new access procedures within 10 days, after which this card no longer will be necessary.

Please contact me with any questions or concerns.

Thank you very much.

# 51 Business Reports

## Purpose, Audience and Media

Business reports communicate facts and sometimes opinions or recommendations based on those facts. Business reports generally describe, in detail, a particular situation. They also can specify information designed to improve the effectiveness of an organization. They should not be confused with federally mandated annual reports (SEC Form 10K) that have very specific requirements (page 57).

The audience for a business report is usually a manager or managers within an organization. Most business reports are internal documents, although some organizations—such as think tanks—prepare reports, often called white papers, for general distribution.

Business reports appear in a variety of media, often depending on the nature and the size of the target audience. Paper, which can be bound together with a staple or binder, remains a common medium. Business reports also can be delivered digitally as PDFs and as interactive websites.

**KEY TO SUCCESS**

Whether formal or informal, good business reports are clear, concise, detailed and well-organized.

## Format/Design

The format of a business report should clarify its organization and aid readability. The text of a business report is single-spaced. Double-space between paragraphs, and do not indent paragraphs. Number the pages of the report, beginning with the first page of the actual text; do not number the title page or the table of contents. The formats of business reports often include the following design and graphics elements:

- Colorful charts, graphs and other visuals to reinforce and clarify meaning. Reports can be long and daunting; graphics can highlight key points and provide visual relief. In online reports, such charts and graphs can be interactive.

- Boldfaced, larger-than-ordinary type for the title on the title page of paper reports and PDFs (18-point Times is a standard size and typeface for report titles)
- Boldfaced type for section titles and subheadlines
- Margins of at least one inch
- White space (extra spacing) between sections. In long reports, new sections often begin on a new page.

As you plan a business report, be sure to review Strategic Design (pages 35–41).

## Content and Organization

Before beginning to write a report, ask yourself these questions:

- What is the purpose of this report? Does everyone involved agree?
- Who is the audience? What does it expect, want and need to learn from this report?

Let the answers to these questions guide the information that the report should contain.

A formal business report contains, in order, the following sections, each of which often begins on a new page. (Less formal versions can discard some of the following sections.)

### *Memo or Letter of Transmittal*

The memo or letter of transmittal typically accompanies paper reports and is paper-clipped to the title page or cover of the report. Such a memo or letter might say, "Attached is the report you requested. If you have questions, please contact me." This page is a matter of courtesy—almost always an important part of strategic communications. When the report is sent via email as an attached file, the email message becomes the memo of transmittal.

### *Title Page*

The title page is often the cover of the report. It includes a title; a descriptive subtitle, if necessary; the name(s) of the author(s); and the date. The title page should clearly communicate that the document is a report on a specific topic. If the title is vague—such as "A Time for Action"—then include a specific subtitle: "Recommendations for Improving Safety in the Central City Factory."

### *Table of Contents*

A table of contents lists each section, in order, and its starting page. (Do not list the span of pages for each section; just list the starting page.) Don't include the memo or letter of transmittal or the title page in the table of contents. With a headline, clearly label this page as the table of contents.

## Table of Charts and Graphs

This optional section lists the names and page numbers of the charts and graphs in the report. This section generally is not listed in the table of contents.

## Executive Summary

An executive summary is a concise, one-page (if possible) overview of the report's highlights. In general, include an executive summary if the report is formal and will take more than 15 minutes to read. Executive summaries highlight the key findings of the report. For example, the executive summary of a problem-solution report would include the most important details of the problem as well as the most highly recommended solutions.

Do not use your executive summary as the introduction to your report. A reader may skip the executive summary. Your report needs a separate introduction (see the next segment). You can avoid this hazard by writing the executive summary last, after you have completed the rest of your report, including its official introduction.

Begin numbering pages with the executive summary.

## Text of the Report

The text generally consists of an introduction, a body (which delivers most of the details) and a conclusion. These three sections follow the organizational cliché of "Tell 'em what you're going to say, say it and tell 'em what you've said."

### Introduction

The introduction states the report's purpose and main point(s). The introduction also establishes the tone of the report (that is, how formal it will be) and the level of language used.

If your introduction seems too brief, consider two things:

- Brevity in a report is an asset as long as it's not confusing or rude.
- If the introduction seems so brief that it's ungraceful, beef it up by announcing the organization of the report. That is, use the introduction to forecast the order of the sections to come. This may repeat some of the table of contents, but such repetition can be useful, reinforcing the report's organization.

## Body

The body delivers the specifics. It explains and develops the main point(s). In most reports, the introduction and conclusion are short; the body is comparatively long.

In many reports, using the subject-restriction-information paragraph-organization technique, when appropriate, can help keep the body focused on the report's subject. Here's how the subject-restriction-information model works: The opening sentence of the paragraph refers to the subject of the report and to the restriction of that subject covered in the paragraph (and, perhaps, in following paragraphs). Specific information then follows. For example, here is a subject-restriction opening sentence: "A second reason for the inefficiency of the Arcadia factory [subject] is obsolete equipment [restriction]." The information following this sentence would prove that the equipment is, indeed, obsolete.

The sections of the body should be organized logically—for example, from the first event in time to the last; from the most important information to the least; from the best solution to the worst; and so forth. Consider using headlines to label sections or chapters in the body.

If the report makes recommendations, present them in the body, not the conclusion. You may wish to announce them in the introduction; you certainly would announce them in an executive summary. In the body, you either can put the evidence first so that when the recommendations appear, they seem logical—or, if there are several recommendations, you might wish to make them one at a time and, after each, provide evidence to support it. A good model is a financial audit report, which first presents evidence, usually a description of an organization's financial practices and financial health. The financial audit report then makes recommendations and, after each recommendation, includes a brief explanation of the problem to be solved or prevented.

## Conclusion

The conclusion reasserts the report's purpose and main point(s) in light of all the information in the body. The conclusion also can note the next action that will or should be taken. If the conclusion seems too short, it can briefly note the most useful sources of information. Remember: Unless brevity seems rude or ungraceful, it is an asset, not a problem.

## Endnotes

Endnotes cite specific sources of information. They're needed only in very formal reports. Each endnote requires a previous marker in the text, usually a small, elevated number like this: [17]

### *List of Works Consulted*

Like endnotes, a list of works consulted is reserved for very formal reports.

### *Appendix*

An appendix would contain articles, tables, charts and other related information. An appendix is optional.

#### BUSINESS REPORT TIPS

1. **Be concise**: Avoid the temptation to pad a report with wordiness, useless information and unneeded sections. Busy people appreciate brevity. Reports should be long enough to thoroughly fulfill their purpose—and no longer.
2. **Get feedback**: If possible, show drafts of your report to your manager as you progress. Reports are important documents, and you should avoid distributing an unedited, unapproved report. Reports reflect not only on the people who write them but also on their supervisors and organizations.
3. **Organize the team**: Long reports often are written by teams. If you are the team leader, assign specific tasks to specific individuals. Assign a deadline for each task. Make these assignments in writing—in a memo—to avoid any misunderstandings. If the report will take several days or weeks to complete, hold quick progress meetings or ask that members of your team send you periodic progress memos.

APPENDIX A

# A Concise Guide to Punctuation

**IMPORTANT! READ THIS:** To understand several rules of punctuation, you must understand what a **clause** is.

A clause is any group of words that has a subject and a verb that has tense. By "a verb that has tense," we mean a verb that sends a time signal, such as past, present or future.

For example, *She wrote a speech* is a clause. It has a subject (*She*) and a verb that has tense (*wrote*)—in this case, past tense. However, *writing the speech* is not a clause. *Writing* can be a verb form, but it doesn't show tense; it's not past, present or future. And *writing the speech* doesn't have a subject, a doer of the action. No one here is doing the writing. Therefore, *writing the speech* is not a clause.

Basically, there are two kinds of clauses: **independent clauses** and **dependent clauses**. An independent clause can stand all by itself as a complete sentence. *She wrote a speech* is an independent clause.

A dependent clause, like all clauses, has a subject and a verb that has tense—but it cannot stand all by itself as a complete sentence. *When I was young* is a clause. However, because it cannot stand all by itself as a complete sentence, it is a dependent clause. A dependent clause must attach itself to an independent clause; otherwise, it becomes an incomplete sentence, also known as a sentence fragment. For example, *When I was young, I studied advertising* is a complete sentence. It consists of a dependent clause attached to an independent clause.

Here's the last part of the grammar lesson. Basically, there are two kinds of dependent clauses: **subordinate clauses** and **relative clauses**. A subordinate clause begins with a subordinate conjunction. The list of subordinate conjunctions includes such words as *because, if* and *when*. *When I was young* is a subordinate clause.

A relative clause begins with a relative pronoun. There are five basic relative pronouns: *that, which, who, whom* and *whose*. In the sentence *The account executive who wrote the proposal won an award*, the words *who wrote the proposal* are a relative clause.

Understand these terms—**clause, independent clause, dependent clause, subordinate clause** and **relative clause**—and you're ready to understand and explain several punctuation rules.

In the rules below, the "PM" simply stands for punctuation mark.

## Commas

**PM1** When a coordinating conjunction (*and, but, or, for, so, nor, yet*) connects two independent clauses, put a comma before the coordinating conjunction.

**Example:** *She wrote the news release, but he distributed it.*

**PM2** In most cases, if *and* (or *or*) is not connecting two independent clauses, do not put a comma before it.

**Example:** *She wrote the news release and put it on the website.*

**Example:** *He writes news releases, video ads and speeches.*

**PM3** Do not connect two independent clauses with only a comma. That error is called a comma splice. (See also Rule PM30.)

**Example:** *She wrote the news release, he distributed it.* (INCORRECT: The comma connects two independent clauses. This is a comma splice.)

**Exception:** A comma can connect two independent clauses when there is a series of at least three independent clauses: *She wrote the news release, he distributed it and they both left to celebrate.*

**Exception:** A comma can attach a question to a previous statement, whether the question is a single word or a complete sentence: *We won, right?* and *We won, didn't we?*

**PM4** Set off an opening subordinate clause with a comma.

**Example:** *If you understand clauses, you'll understand more about punctuation.*

**Tip:** The list of subordinate conjunctions includes *if, when, although, though, because, while, unless, as soon as, before* and *after*.

**PM5** Do not (usually) put a comma before a subordinate clause that ends a sentence. The comma isn't wrong; it's optional and usually not used.

**Example:** *You'll understand more about punctuation if you understand clauses.*

**PM6** If a relative clause narrows down the meaning of the noun that comes before it, do not set it off by commas.

**Example:** *The account executive who wrote the proposal won an award.*

**Explanation:** The relative clause *who wrote the proposal* narrows down what we mean by the noun *executive*. In other words, the clause is essential to what we mean by *executive*. We don't mean just any executive; we mean the one who wrote the proposal.

**Tip:** Sometimes the relative pronoun and its verb are deleted from the sentence: *My favorite song sung by Lenny Kravitz is "Fields of Joy."* (In this sentence, *that is sung by Lenny Kravitz* has been reduced to *sung by Lenny Kravitz*.)

**PM7** If a relative clause does not narrow down the meaning of the noun that comes before it, set it off with commas.

**Example:** *My biological father, who lives in London, works in corporate communications.*

**Explanation:** The relative clause *who lives in London* does not narrow down what we mean by the noun phrase *biological father*. The speaker has only one biological father.

**Tip:** Sometimes the relative pronoun and its verb are deleted from the sentence: *My favorite song, sung by Lenny Kravitz, is "Fields of Joy."* (In this sentence, *which is sung by Lenny Kravitz* has been reduced to *sung by Lenny Kravitz*. This differs from the tip in Rule PM6. In that guideline, the song is the speaker's favorite just among Kravitz's songs. In this tip, for Rule PM7, the song is the speaker's favorite song among all songs.)

**PM8** Use commas to set off nouns and noun phrases when they immediately follow a noun or noun phrase that means the same thing.

**Example:** *Julie Smith, our new president, will address the stockholders.*

**Example:** *Our new president, Julie Smith, will address the stockholders.*

**Tip:** If you can drop the second noun (or noun phrase) from the sentence and the sentence still works grammatically, set the second noun (or noun phrase) off by commas.

**PM9** Do not put a comma between a title and a name when you can substitute *Mr.* or *Ms.* for the title.

**Example:** *President Julie Smith will address the stockholders.* (Ms. Julie Smith will address the stockholders.)

**PM10** Do not set off a noun (or noun phrase) with commas when it narrows down the meaning of a preceding noun.

**Example:** *My associate Arnold Jones will address the stockholders.* (The noun *Arnold Jones* narrows down the noun *associate;* it tells which associate.)

**Example:** *Our newsletter* Employees Today *just won a national award.* (This is accurate only if the company has more than one newsletter. In that case, *Employees Today* narrows down the noun *newsletter*; it tells which of your organization's newsletters won the award.)

**PM11** Set the number designating a year off with commas when the number follows a month and a date. When a date follows a day of the week, set the date off with commas.

**Example:** *Jan. 1, 2020, was a memorable day. Wednesday, Jan. 1, was a memorable day.*

**Example:** *I remember Jan. 1, 2020. I remember Wednesday, Jan. 1.*

**Tip:** Most stylebooks agree that when only the month and year are specified, commas are unnecessary:

*January 2020 was a profitable month for our company.*

**PM12** Set off state and country names with commas when they follow city names.

**Example:** *Weslaco, Texas, is near Mexico.*

**Example:** *She has worked in Asolo, Italy, for five years.*

**PM13** Use a comma to separate adjectives that modify a following noun. (If you can substitute *and* for the comma, the comma usage is correct.)

**Example:** *The direct-mail campaign was an effective, timely tactic.*

**Exception:** *He wrote an excellent annual report.* (Do not put a comma between the adjectives *excellent* and *annual* because *annual* is considered to be part of the noun phrase *annual report. Excellent* modifies the entire noun phrase *annual report.*)

**PM14** Do not use a comma before the final item in a series. (Stylebooks differ on this point. The Associated Press prefers no comma in this situation.)

**Example:** *The campaign includes a video ad, an audio ad and a magazine ad.*

**Exception:** Even the Associated Press recommends inserting a comma when an item in a series includes an *and*: *The campaign includes a news release, a speech, and print and video ads.*

**Exception:** When the items in a series are wordy and highly detailed, the Associated Press recommends including a comma before the final *and*.

**PM15** Use a comma before opening quotation marks when both of two things occur: (a) the quotation immediately follows *said* (or a similar word) and (b) when the quotation answers the question *said what?*

**Example:** *She said, "Our annual report is a stunning example of computer graphics."*

**Example:** *She said that the annual report was "a stunning example of computer graphics."* (In this second example, do not put a comma before the opening quotation marks. The quotation does not immediately follow *said*. Also, the quotation by itself does not answer the question *said what?* The quoted passage needs the help of the words *the annual report was* to answer that question.)

**Tip:** Do not automatically put a comma before quotation marks. Follow Rule PM15.

**PM16** When the attribution (the *she said*) follows a quotation or a paraphrase, set the attribution off with a comma.

**Example:** *"Our annual report is a stunning example of computer graphics," she said.* (Remember that the comma goes inside the closing quotation marks. See Rule PM20.)

**Example:** *The company's annual report is superb, she said.*

**PM17** When an attribution (the *she said*) follows a quotation that ends with a question mark or an exclamation point, do not include a comma.

**Example:** *"Did she praise the annual report?" he asked.*

**PM18** When you interrupt a quotation with an attribution (the *she said*), put a comma before it. If the attribution comes at the end of a sentence, put a period after it. If it does not come at the end of a sentence, put a comma after it.

**Example:** *"Our annual report is superb," she said. "I love the computer graphics."*

**Example:** *"Our annual report is superb," she said, "but we'll do even better next year."*

**PM19** Do not put a comma after an attribution (the *she said*) that introduces a paraphrased quotation.

**Example:** *She said that the annual report was superb.*

**PM20** When a comma is next to a closing quotation mark, put the comma inside the quotation—even if that seems to make the comma part of a title.

**Example:** *"I'll address the stockholders," she said.*

**Example:** *His favorite song, "Fields of Joy," sung by Lenny Kravitz, is on my playlist.* (Why is there a comma before the song title? See Rule PM8. Why is *sung by Lenny Kravitz* set off by commas? See Rule PM7.)

**PM21** When someone is directly addressed by name (or a substitute for the name), set that name off by commas. (This is called "direct address" because, of course, you're directly addressing someone.)

**Example:** *Good morning, class.*

**Example:** *I think, Mr. Jones, that you'd better sit down.*

**PM22** When the word *yes* or the word *no* is used to answer someone, set it off with commas.

**Example:** *"Yes, our annual report is superb," she said.*

**Example:** *"But, no, I must disagree," he said.*

**PM23** When someone's age follows their name, set the age off by commas.

**Example:** *Julie Smith, 47, is our new president.*

**PM24** When an *-ing* phrase (a participial phrase) modifies the subject of a sentence but is not essential to the meaning of the subject (meaning the *-ing* phrase could be dropped), set the *-ing* phrase off by commas.

**Example:** *Finishing the annual report, he laughed with delight.*

**Example:** *He laughed with delight, waving the annual report excitedly.*

**Example:** *He ran into the office and, waving the annual report excitedly, began to shout.*

**PM25** Rule PM24 also applies to phrases beginning with a past participle, most of which end in *-ed*. When such a phrase modifies the subject of a sentence but is not essential to the meaning of the subject, set it off by commas.

**Example:** *Finished with the annual report, he laughed with delight.*

**Example:** *He slumped in exhaustion, finished at last with the annual report.*

**Example:** *He rose and, finished at last with the annual report, shouted with joy.*

**PM26** When a phrase of four or more words opens a sentence, set it off by a comma. Also, set off a shorter opening phrase with a comma if doing so reduces ambiguity.

**Example:** *In a little more than three hours, he finished writing the sales letter.*

**Tip:** This guideline is a matter of style, not a grammatical rule.

Ask your professor or boss what their policy is.

**Example:** *Ever since, I was afraid of cats.* (This comma reduces ambiguity.)

**PM27** When an adverb (most end in *-ly*) begins a sentence, set it off with a comma.

**Example:** *Generally, direct-mail campaigns are inexpensive.*

**PM28** When the word *not* introduces a contrary idea in a sentence, put a comma before *not.*

**Example:** *The annual report was successful, not at all a failure.*

**PM29** Set interjections off by commas. Interjections include words such as *um* and *well* and phrases such as *by the way* and *you know.*

**Example:** *Our annual report was, well, a little too expensive.*

**PM30** Set off by commas the so-called conjunctive adverbs: *however, though, therefore, furthermore, consequently, moreover* and similar words. Be careful, however, to avoid comma splices (Rule PM3).

**Example:** *Therefore, he proposed an advertising campaign.*

**Example:** *Our sales brochures, consequently, have been effective.*

**Example:** *Our sales brochures are effective; however, our telemarketing needs revision.* (Note how a semicolon is used here to avoid a comma splice. See Rules PM3 and PM33.)

**PM31** Do not set off an opening coordinating conjunction with a comma.

**Example:** *And, we never saw her again.* (INCORRECT)

**Example:** *But, that's for you to decide.* (INCORRECT)

**Exception:** The comma is correct when it works with a following comma to set off a phrase: *And, after that, we never saw her again.*

**PM32** Do not separate subject from verb with a single (unpaired) comma.

**Example:** *Her habit of always being right and letting everyone know it, is annoying.* (INCORRECT)

## Semicolons

**PM33** Use a semicolon to connect two closely related independent clauses when a period seems to be too harsh a separation.

**Example:** *He likes our current strategy; she distrusts it.*

**PM34** Use a semicolon to separate different items in a series in which all or some of the items include commas.

**Example:** *We'll revise our media kits, which have been too expensive; our direct-mail packages, which haven't been well-targeted; and our print ads, which haven't included enough selling points.*

## Colons

**PM35** Use a colon only after an independent clause. (In other words, an independent clause must come before a colon.) Note: The sentence must end with the word(s) introduced by the colon; the sentence cannot continue after that.

**Example:** *She has one great strength: punctuation.*

**Tip:** When a complete sentence follows a colon, most stylebooks recommend beginning that sentence with a capital letter.

**Exception:** In the title of a document, a colon need not follow an independent clause—*Our Future: A Roadmap*.

**PM36** Do not put a colon before a list unless the colon follows an independent clause.

**Example:** *Her strengths include: grammar, punctuation and spelling*. (INCORRECT: The colon does not follow an independent clause.)

**Example:** *Her strengths include grammar, spelling and punctuation*. (No punctuation follows *include*.)

**Exception:** Sometimes an incomplete sentence introduces a bulleted list.
A colon at the end of the incomplete sentence has become traditional. Ask your professor or boss what their policy is.

*In our office, we produce:*
*newsletters;*
*policy and procedure documents; and*
*annual reports*

## Quotation Marks

**PM37** When a quotation appears within a quotation, use single quotation marks within the original double quotation marks.

**Example:** *"I love the song 'Fields of Joy,'" she said.*

**Tip:** When a comma or a period is next to a single closing quotation mark, it will go inside the mark.

**PM38** When a quotation continues beyond one paragraph, begin each new paragraph with opening quotation marks. Do not insert closing quotation marks until the quotation has concluded.

**Example:** *"Our annual report is superb," she said. "I love the computer graphics.*
*"Our direct-mail packages also have been effective. Our new databases have helped significantly.*
*"Finally, our advertising campaigns have been good, but we must try to target them more effectively."*

**PM39** When quotation marks appear in a headline, use single quotation marks to save space.

**Example:** *CEO predicts 'brilliant future'* (headline)

**PM40** Quotation marks sometimes are used to signify titles of creative works such as books, songs and movies. Some stylebooks recommend italics for those situations. Ask your professor or boss what their policy is.

**Example:** *My favorite old Beatles song is "In My Life."*

**Tip:** The Associated Press recommends using quotation marks for the titles of books (except the Bible and reference books such as catalogs and dictionaries); computer games (but not software); movies; operas; plays; poems; songs; TV shows; lectures; speeches; and works of art.

**PM41** Put quotation marks around unfamiliar words.

**Example:** *In his speech, he used a stylistic device called an "antimetabole."*

## Question Marks

**PM42** Put a question mark inside closing quotation marks when the quotation is a question. When the quotation is not a question, put the question mark outside the quotation.

**Example:** *"Do you like our annual report?" she asked.*

**Example:** *"Did he say, 'I love your annual report'?"*

**Tip:** See also Rule PM17.

**PM43** When a sentence culminates in two questions, use only one question mark.

**Example:** *"Did he ask, 'Where is your annual report?'"*

**PM44** Realize that although a question mark can end a sentence (like a period), it also can function like a comma and allow the sentence to keep going.

**Example:** *"Do you like our annual report?" she asked.*

**Example:** *If he were to ask, "Was your annual report effective?" I know what I would answer.*

**PM45** Do not put a question mark at the end of a paraphrased question that actually is a statement.

**Example:** *She asked how they planned to finish the annual report.*

## Exclamation Points

**PM46** Put an exclamation point inside closing quotation marks when the quotation is an exclamation. When the quotation is not an exclamation, put the exclamation point outside the quotation.

**Example:** *"I love this annual report!" she shouted.*

**Example:** *I'm devastated that he called the annual report "mediocre"!*

**Tip:** See also Rule PM17.

**Tip:** Journalists dislike exclamation points in news stories.

Avoid them in news releases and media kits.

**PM47** Realize that although an exclamation point can end a sentence (like a period), it also can function like a comma and allow the sentence to keep going.

**Example:** *"I love this annual report!" she shouted.*

**Example:** *If he were to shout, "I love this annual report!" would you be surprised?*

## Periods

**PM48** When a period is next to a closing quotation mark, put the period inside the quotation—even if that seems to make the period part of a title.

**Example:** *She said, "This annual report is superb."*

**Example:** *He said, "My favorite song is 'Fields of Joy.'"*

**Exception:** When a parenthetical phrase follows a quotation at the end of a sentence, put the period after the parenthetical phrase: *Our annual report includes a page titled "Projections for the Next Decade" (p. 47).*

**PM49** When an abbreviation with a period ends a sentence, do not add a second period.

**Example:** *Bert lives at 123 Sesame St.*

**Tip:** A period that follows an abbreviation does not necessarily end a sentence. *Janet Smith, Ph.D., is our new CEO.*

**PM50** When words in parentheses close a sentence, the period goes inside the parentheses only if the entire sentence is in parentheses. If the beginning of the sentence is not in parentheses, the period goes outside.

**Example:** *(We distributed that PSA in August.)*

**Example:** *We distributed that PSA this past summer (in August).*

## Dashes

**PM51** Use dashes to set off a sudden thought that interrupts the progress of a sentence. (A dash generally is typed as two hyphens.)

**Example:** *Our annual report—did you know it won a national award?—is superb.*

**PM52** Use a dash to create a dramatic pause in a sentence.

**Example:** *Our annual report will win a national award—if we finish it on time.*

**PM53** Use dashes to set off the expansion of a word or concept when that expansion interrupts the progress of the sentence.

**Example:** *The qualities that made our annual report a winner—conciseness, accuracy and infographics—characterize all of our investor relations publications.*

**PM54** Use a dash as an informal substitute for a colon. (See Rule PM35.)

**Example:** *She has one great strength—punctuation.*

## Hyphens

**PM55** Use a hyphen to connect two or more modifiers when (a) they act as one word to modify a following noun and (b) when the absence of the hyphen could lead to confusion.

**Example:** *Our easy-to-assemble products need no instructions.*

**Exception:** The Associated Press recommends not using a hyphen with *very* compounds and compounds that involve adverbs that end in *-ly*.
**Exception:** The Associated Press recommends not using a hyphen when a compound modifier before a noun is well-known and does not need a hyphen for clarity.

**PM56** Use a hyphen when a prefix might cause confusion by creating a word identical to—but with a meaning different from—the word you want.

**Example:** *His re-creation of the situation was impressive.*

**PM57** When two or more words work separately with another word and those combinations come before and modify yet another word, use suspensive hyphenation.

**Example:** *Please prepare 15-, 30- and 60-second commercials.*

## Apostrophes

**PM58** Use apostrophes in contractions to indicate where a letter or letters have been deleted.

**Example:** *We're proud of our annual report.*

**Tip:** Remember that the word *it's* (with an apostrophe) is the contraction of *it is*. Without the apostrophe, *its* is a possessive personal pronoun: *The company introduced its new CEO.*

**PM59** Use an apostrophe to indicate possession.

**Example:** *The company's outlook is excellent.* (singular common noun)

**Example:** *The officers' spouses will join them after the meeting.* (plural common noun)

**Example:** *The boss's word is law.* (singular noun ending in *s*)

**Example:** *Davis' report was impressive.* (singular name ending in *s*)

**Tip:** Rules and style guidelines for using apostrophes are extensive and complex. For a detailed list, see the Associated Press Stylebook. Also, ask your professor or boss what their policies are.

**PM60** Do not use an apostrophe with possessive personal pronouns.

**Example:** *You believe that the credit is yours; however, your associates believe it is theirs.*

**Tip:** The possessive personal pronouns include *yours, his, hers, its, ours* and *theirs.*

**PM61** Do not use an apostrophe when a plural noun acts as an adjective for a following noun.

**Example:** *The conference will be in the teachers lounge.*

**Tip:** A word like *teachers* will not be possessive (and therefore will not have an apostrophe) if you can reverse the phrase and add the word *for*, as in a lounge for teachers. This shows that the lounge does not belong to teachers; rather, the lounge is for them.

**PM62** Use an apostrophe with decade abbreviations.

**Example:** *That strategy was successful during the '90s.*

## Parentheses

**PM63** Use parentheses to set off comments or citations that are grammatically nonessential.

**Example:** *A successful document (such as our annual report) is the result of extensive planning.*

**Tip:** Don't overuse parentheses. Dashes can often substitute for them.

**Tip:** See Rule PM50 for period placement and parentheses.

# APPENDIX B A Concise Guide to Grammar

Grammar may not be your favorite subject. In fact, grammar may not be in your Top 100 favorite subjects. However, strategic writers must have a solid command of grammar. Great ideas sometimes don't survive bad grammar. Also, a shaky knowledge of grammar can cause strategic writers to avoid problems by avoiding certain sentence structures that might otherwise be effective.

A second reason to improve your knowledge of grammar is to improve your career prospects. When you know what is correct grammatically and can explain it to others, you have a rare, marketable talent.

In this brief section, the authors describe the seven most common grammatical errors we find in student (and, sometimes, professional) writing. We've numbered the items so that you and your professor can easily refer to them. For additional information on good grammar, please see Appendix A: A Concise Guide to Punctuation. That appendix includes punctuation-related grammatical errors such as the comma splice.

## G1 Pronoun Disagreement

In general, don't replace a singular noun with a plural pronoun. For example, don't write *That company loves their employees*. The noun *company* is singular, but the pronoun that later replaces it (*their*) is plural. Avoid a singular-to-plural shift like that, especially when the noun does not refer to a specific person (see the final paragraph of G1).

This form of pronoun disagreement has two basic solutions: Either make both terms plural or make both terms singular. If both terms are plural, your sentence could be something like *That company's leaders love their employees*. Now we move from a plural noun (*leaders*) to a plural pronoun (*their*). If both terms are singular, your sentence could be something like *That company loves its employees*.

A gender-neutral first term—such as *employee*—can sometimes lead to an awkward *his/her* decision later in the sentence: *Each employee should take his or her phone . . . .* If possible, make both terms plural: *All employees should take their phones*. However, the Associated Press now uses *they* as a gender-neutral singular pronoun in the following circumstances:

- When the *he/she* alternative is awkward and substituting a plural noun doesn't work well.
- When an individual requests that the pronouns *they/them/their* be used to describe that individual. (In such circumstances, the Associated Press recommends explaining to readers that the individual prefers a gender-neutral pronoun.)

Strategic writers should consult their editors to determine policy in this evolving area.

## G2 False Series/Lack of Parallelism

This problem occurs in sentences that include a series, such as *Strategic writing requires research, creativity and diligence*. In that correct sentence, the series consists of three nouns: *research, creativity* and *diligence*. A false series—also known as a lack of parallelism—occurs when different items in the series become different parts of speech (such as a mix of nouns and verbs). The breakdown often comes in the last item in the series. For example, this sentence has a false series or a lack of parallelism: *Strategic writing requires research, creativity and to work hard*. In that sentence, our series is noun, noun, verb—an ungrammatical lack of parallelism.

Probably the most common form of false series happens like this: *Our employees value hard work, dedication and they especially value honesty*. In this sentence, our series is noun, noun, complete sentence. To fix the problem, we need to downsize our series from three items to two items and add a new clause: *Our employees value hard work and dedication, and they especially value honesty*. Or *Our employees value hard work, dedication and especially honesty*.

## G3 Subject-Verb Disagreement

Verbs have to reflect the number of the subject. In other words, a verb has to reflect whether its subject is singular or plural. It often are easy to tell when you makes a mistake in subject-verb agreement (as this sentence illustrates). However, when the verb is a long way away from its subject—or when we're not sure whether the subject is singular or plural—subject-verb disagreement can occur.

Consider this sentence: *A herd of wombats are stampeding in the streets*. Disagreement? Yes, because *herd* is our subject, and in American English (though not in British English), *herd* is singular. In American English, collective nouns such as *herd, group, jury* and *team* are singular (in British English, they're plural). So unless you're having a tea party with Princess Charlotte of Wales, you'd say *A herd of wombats is stampeding in the streets*. (In this sentence, *wombats* is not our subject. *Wombats* is the object of a preposition; therefore, it can't be our subject.)

Now try this sentence: *His ability to analyze problems and develop solutions make him an outstanding leader*. That verb should be *makes,* not *make*. The verb's subject is *ability*—but the subject is so far away from the verb that we can lose sight of the connection. Don't let distance lead you into subject-verb disagreement.

Here's another tough one: *My talent plus your ambition are a good combination*. Disagreement? Yes, because *plus* doesn't equal *and*. *Plus* doesn't give us a plural subject. The correct version would be *My talent plus your ambition is a good combination*—an ugly sentence that you probably should revise. The same principle applies to *as well as*: That phrase doesn't equal *and*. It doesn't give us a plural subject. We would have subject-verb disagreement if we were to write *The CEO as well as the CFO are in the room*. Unfortunately, it should be *The CEO as well as the CFO is in the room*—again, an ugly sentence that you probably should revise: *The CEO and the CFO are in the room*.

One last disagreement: *Either the three vice presidents or the CEO have leaked this information to the news media*. When you have a two-part subject united by the word *or,* whichever part of the subject sits closest to the verb controls the verb. In this case, *CEO* sits closest to the verb—and so the verb should show a singular subject: *Either the three vice presidents or the CEO has leaked this information to the news media*. Once again, this is an ugly sentence that you probably should revise.

## G4 *Who* and *Whom*

Believe it or not, there's an easy shortcut for this one—but first, let's do the basic grammar. (Sorry: The authors are professors, after all.) *Who* and *whom* are pronouns. *Who* is a subject pronoun (a so-called nominative pronoun). Because of that, *who* almost always will be the subject of a verb that has tense (such as past, present or future). *Whom* is an objective pronoun. Because of that, *whom* can never be the subject of a verb. Instead, *whom* will always be an object, such as a direct object or the object of a preposition.

That knowledge gives us one solution to the *who/whom* problem: If the word is the subject of a verb, we know it must be *who*. If it's not the subject of a verb, we know it must be *whom*.

Now for that shortcut, also known as the chop method. Because *who* is a subject pronoun, it is grammatically equal to *he*. Because *whom* is an objective pronoun, it is grammatically equal to *him*. To use the chop method,

1. Read the sentence in question and stop right before the *who/whom* word.
2. Now throw away that first part of the sentence.
3. Mentally substitute *he* and *him,* instead of *who* or *whom,* in the next several words that follow. In other words, place *he* and *him* in the clause introduced by the *who-whom* decision. Remember that we define *clause* on page 333.
4. If *he* works, you need *who*. If *him* works, you need *whom*.

For example, try the chop method on this sentence: *Give it to whoever/whomever wants it.*

1. Give it to whoever/whomever wants it.
2. whoever/whomever wants it.
3. *He wants it* or *Him wants it*. (Clearly, it's *He wants it*.)
4. The answer is *Give it to whoever wants it*.

Try it again on this sentence: We'll award the contract to whoever/whomever we trust.

1. We'll award the contract to whoever/whomever we trust.
2. whoever/whomever we trust.
3. *He we trust* or *Him we trust*—or, better, We trust him. (Clearly, it's *We trust him*.)
4. The answer is *We'll award the contract to whomever we trust*.

With practice, you can learn to do the chop method quickly and accurately.

## G5 Dangling Participles, Dangling Modifiers

This one sounds awful, but it's not too hard to understand. Basically, two kinds of participles exist (don't ask us why): present participles and past participles. Present participles are verb forms that end in *-ing: Running, praising, screaming* and *writing* all can be present participles. Past participles are verb forms that complete this passage: *I have* _____. The words *run, praised, screamed* and *written* all can be past participles.

Here comes the rule: When these participles introduce or help introduce a phrase that opens a sentence, that phrase modifies the subject. In other words, the phrase modifies the first noun or pronoun to follow the phrase. A dangling participle or dangling modifier occurs when the wrong word or words become the subject. For example, here's a classic dangling modifier from English 101: *Flying over the North Pole, an iceberg was seen*. We all know what the writer meant—but tough luck. *Flying over the North Pole* is an opening participial phrase; therefore, it modifies *iceberg*. Hence, the amazing flying iceberg.

Here's another one: *Screamed into the wind, he knew his words were lost*. Again, we know what the writer meant. But *Screamed into the wind* is an opening participial phrase; therefore, it modifies *he*. Hence, he—not his words—was screamed into the wind.

Not all danglers are participles. See the problem in the following sentence? *As editor of the annual report, your knowledge of grammar is essential*. Technically, *your knowledge of grammar* is not the *editor of the annual report*. In this case, we have a dangling modifier.

There's no easy way to fix a dangling modifier. You have to tear down the sentence and try again. Solutions to the dangling participles could be *Flying over the North Pole, we saw an iceberg* and *He knew that his words, screamed into the wind, were lost.* A solution to the dangling modifier could be *Because you are the annual report's editor, your knowledge of grammar is essential.*

## G6 *That* and *Which*

*That* and *which* are versatile words with many duties. But when they introduce clauses (see page 333), *which* sometimes gets confused with *that*, which leads to a minor grammatical error. (PM6 and PM7 on pages 334–335 cover a similar concept.)

Use *that* to introduce a group of words (a clause) that is essential to the meaning of the immediately previous noun. In other words, use *that* when the clause narrows down, or restricts, the meaning of the previous noun: *The horse that won the race has been stolen.*

In the previous sentence, not just any horse has been stolen—it's the horse *that won the race*. That information is essential; it restricts and narrows down which horse we mean. In the same situation, we would not write *The horse which won the race has been stolen*. With clauses, *which* introduces nonessential, nonrestrictive information. Also, you should always set a *which* clause off with a comma or commas: *This grammar section, which we hope you enjoy, is getting pretty long.* In the previous sentence, the *which* clause doesn't narrow down what we mean by *This grammar section*. The clause is nonessential, so it uses *which* and is set off with commas. Sometimes, a *which* clause closes a sentence and requires only one comma: *She loves the movie "Casablanca," which I've seen many times.*

## G7 *I* and *Me*

*I* and *me* can seem like the city kid and the country cousin. *I* seems distinguished and refined; *me* can seem like a hick. *I* seems like proper grammar, as in *King Henry and I will dine now, Jeeves. Me* can be an embarrassment, as in *Me and him are fixin' to chew some tobacco.* (In case we sound prejudiced against country folks, much of this book was written in Kansas and North Carolina.)

If you're lucky, you don't suffer from this weird distrust of *me*. If you're unlucky, get over it. *Me* is a perfectly good word; just don't use it in the wrong place (as in the tobacco sentence).

The difference between *I* and *me* is the difference between *who* and *whom*, discussed in G4. *I* almost always will be the subject of a verb that has tense (such as past, present or future). *Me* is an objective pronoun. Because of that, *me* can never be the subject of a verb. Instead, *me* will always be an object, such as a direct object or the object of a preposition.

Probably the most common *I*/*me* error is the avoidance of *me* as an indirect object: *The CEO told Dave and I that we would be promoted.* You can avoid this problem by testing the sentence without Dave. You wouldn't say *The CEO told I that I would be promoted*.

A second common error is the avoidance of *me* in preposition phrases: *Our new chief financial officer sat at the table between Marion and I*. You probably can spot the problem: *I* needs to be the subject of a verb, and there's no verb that lacks a subject. We need to use *me* because the word is the object of the preposition *between*. Objects need to be objective pronouns.

APPENDIX C

# A Concise Guide to Style

Sorry, no fashion tips here. This brief appendix focuses on writing style. Like grammar, style provides guidelines for good writing. Unlike grammar, however, style doesn't consist of rules that everyone accepts. For example, we all can agree that *Her and me don't write so good* is bad grammar. But what about something like a state name? Should *California* be abbreviated? If so, should we use *Calif.*? Or *Cal.*? Or *CA*? Grammar doesn't cover such matters. That's why we need style.

Organizations need consistent writing style to avoid the sloppy appearance of stylistic disagreements among their strategic writers. Imagine, for example, four different strategic writers working for the same website. Unless they use the same style for things such as state names, numbers and company names, their editor will spend needless hours enforcing consistency. Any inconsistencies that escape the editor and appear in print can make the organization look careless.

Newspaper journalists were among the first groups to adopt so-called stylebooks—that is, books that specify the style an organization will use. Many newspapers in the United States use the Associated Press Stylebook (more commonly called the AP Stylebook). Because that manual is well-known, well-organized and easy to use, many organizations have adopted the AP Stylebook for their own written communications. The style tips that follow are just a few of the hundreds of guidelines in that book. If you don't already have a copy of the AP Stylebook, we recommend that you get one. We've numbered the tips below so that you and your professor can easily refer to them. In each entry, we supply only the basics. The AP Stylebook goes into greater detail.

## S1 Business Titles

Your organization may wish to overrule AP style for business titles. AP recommends capitalizing a person's business title only when it comes immediately before the person's name *and* you can substitute the word *Mr.* or *Ms.* for the title: *Chief Executive Officer Alvin Fernald.* In all other cases, AP recommends lowercase letters for business titles: *Alvin Fernald, chief executive officer, will address the board of directors. The chief financial officer also will speak. The director of investor relations, Esther Summerson, will attend.*

In news releases and other documents sent to journalists, you should follow this style (again, many journalists use AP style). But for communications within your own organization, you might be wise to capitalize all titles all the time. Otherwise, you'll constantly explain to Employee A why his title *wasn't* capitalized when Employee B's title *was*. And even after you explain AP style to Employee A, he'll say, "OK, I understand. But you still capitalized *her* title and not *mine*!"

AP style isn't like grammar. You can overrule it. Just be sure everyone on your communications team knows and agrees.

## S2 Abbreviations in Company Names

In formal company names, abbreviate the following words:

| | |
|---|---|
| Brothers: | Bros. |
| Company: | Co. |
| Companies: | Cos. |
| Corporation: | Corp. (when used at the end of a company's name) |
| Incorporated: | Inc. (Do not put a comma before or after *Inc.*) |

Examples of AP style include *Smith Bros. Co.* and *Bingo Corp. Inc.*

## S3 Second Reference

When you use a generic term for a second reference to an organization, lowercase that term: *The Flora Family Foundation donated $2 million. Last year, the foundation donated $1 million.*

When a second reference uses the initials of an organization, do not put those initials in parentheses after the first reference: *The Flora Family Foundation donated $2 million. Last year, the FFF donated $1 million.*

## S4 Numbers

Spell out the numbers zero through nine. Use figures for 10 and above. When you get to millions and above, use figures and words—for example, *1 million; 4 billion; 12 trillion.*

AP lists several exceptions to this policy. Use figures instead of words for ages, money, percentages, room numbers, temperatures (except zero) and weights. AP lists more exceptions in its *numerals* entry.

For fractions smaller than one, spell out the amount and use a hyphen: *three-fourths.* For numbers larger than one, use a decimal point and figures (not words) when possible:

*1.75*. Don't extend numbers past hundredths. In other words, don't include more than two numbers to the right of the decimal point.

## S5 Abbreviation of Month Names

When the name of a month appears with a date, as in *Sept. 1*, use these abbreviations: *Jan., Feb., Aug., Sept., Oct., Nov.* and *Dec.* Spell out the names of the other months. When the name of a month stands alone without an accompanying date, always spell out the full name.

## S6 State Names

AP no longer abbreviates state names when they follow city names in sentences. Spell out the full name of the state in those instances. AP does abbreviate most state names when they appear in datelines at the beginning of news stories. In those dateline abbreviations, AP does not use postal codes; instead, it uses the abbreviations listed below. In datelines, AP recommends spelling out *Hawaii* and *Alaska* and spelling out state names of five or fewer letters: *Idaho, Iowa, Maine, Ohio, Texas* and *Utah*. In its *Datelines* entry, AP lists cities—such as Boston and Miami—that don't require state-name abbreviations.

Alabama: Ala.
Arizona: Ariz.
Arkansas: Ark.
California: Calif.
Colorado: Colo.
Connecticut: Conn.
Delaware: Del.
Florida: Fla.
Georgia: Ga.
Illinois: Ill.
Indiana: Ind.
Kansas: Kan.
Kentucky: Ky.
Louisiana: La.
Maryland: Md.
Massachusetts: Mass.
Michigan: Mich.
Minnesota: Minn.
Mississippi: Miss.
Missouri: Mo.
Montana: Mont.
Nebraska: Neb.
Nevada: Nev.
New Hampshire: N.H.
New Jersey: N.J.
New Mexico: N.M.
New York: N.Y.
North Carolina: N.C.
North Dakota: N.D.
Oklahoma: Okla.
Oregon: Ore.
Pennsylvania: Pa.
Rhode Island: R.I.
South Carolina: S.C.
South Dakota: S.D.
Tennessee: Tenn.
Vermont: Vt.
Virginia: Va.
Washington: Wash.
West Virginia: W.Va.
Wisconsin: Wis.
Wyoming: Wyo.

## S7 Percentages

As was the case with state names, AP style has evolved when it comes to the use of percentages. Use the % symbol, not the word *percent,* with numerals: 8%, 65%, 500%. Do not put a space between the numeral and the symbol. As noted here in S4, use numerals, not words, for percentages.

## S8 Commas in a Series

Use a comma to separate items in a series, but do not use a comma before the final *and* or *or*: *We need more computers, printers and scanners.*

Do use a comma before the final *and* or *or* when elements in the series use *and* or *or*: We invited three couples: *Carrie and Curtis, Claudia and Wayne, and Kay and Duncan.*

Whether your organization uses the AP Stylebook or a different book—or even composes its own guidelines—your organization should have a stylebook. Inconsistent or illogical style can distract a reader and lessen the impact of your document's strategic message.

APPENDIX D

# Tips for Oral Presentations

Strategic writers sometimes must stand up and present their work to a group. For example, you might help explain a new ad campaign to a client's management team. Or you might lead the presentation of a proposal for a new marketing strategy.

Confidence in oral presentations comes from experience. But until you gain that experience, you might be uncomfortable speaking in front of groups—many of us are. This short chapter offers tips for making successful oral presentations. We recommend a four-step process that works for all forms of strategic communication: research, planning, communication and evaluation.

## Research

In the earlier sections of this book, each segment that describes a document begins with an analysis of purpose, audience and media. You need to study those three areas before you make your presentation.

In researching your purpose,

1. Identify the main reason for the presentation. What do you hope to accomplish?
2. Pinpoint your strategic message. What is the one main point of your presentation that all your information will support?
3. Find the information that best supports your strategic message.

In researching your audience members,

1. Learn who they are. What unites them? Why are they a group?
2. Learn what they hope to gain from your presentation.
3. Identify the leaders and decision makers. You'll want to devote extra attention to them.
4. Determine how long you're supposed to speak. What is your time limit? Do audience members hope to ask questions when you're done?

In researching your media,

1. Determine what technology will be available. Will you be able to use computer projection? Will you need your own laptop?
2. Learn about the layout of the presentation room. Will you stand at the head of a table? Or will you present from a lectern in an auditorium? Will you need a microphone? Will a water bottle be available?

## Planning

Begin by planning to be yourself. If you try to act like someone else during the presentation, you carry the double burden of acting and presenting. Just be yourself at your best. In addition,

1. Consider how to combine your strategic message with your audience's self-interest. Audience members want to hear about themselves and what your message means to them.
2. Write an outline for your presentation. (Write a script, if necessary, but remember: You'll need to maintain eye contact with audience members.)
3. Consider writing a brief introduction—a biography of yourself—for the person who will introduce you.
4. Think twice about beginning with a joke. Professionals often discourage such a beginning. If the joke falls flat, your presentation has a shaky start.
5. Consider this traditional beginning: After you have been introduced, thank the audience. Pause for just a moment. Smile. Look at the people in the room. Show them that you're confident, ready and excited about the information you're about to deliver. Don't rush. When audience members see that you're relaxed, they'll relax.
6. Consider using visual aids to emphasize your main points. Studies show that well-designed visual aids increase an audience's comprehension of your information. If you use such aids, keep them simple. You don't want audience members reading while you're talking.
7. Plan a presentation that consumes about three-fourths of the allotted time. Allow for surprises and questions. Know where you can cut your presentation if necessary.
8. Practice your presentation—first alone and then in front of colleagues who will evaluate your performance. No, that's not fun, but it's necessary. Better to learn about errors during trial runs than during the real performance.
9. If possible, practice in the room where you will deliver the presentation. If that's not possible, practice in a similar room.
10. During your final practices, wear the clothing that you'll wear during the presentation—usually clothing at or just above the level of formality of your audience's attire.

11. Finally, plan for problems. What will you do if the computer projection system fails? (Will a document projector, such as an Elmo, be available? Should you bring handouts?) What will you do if you lose your outline or script? (Have a second copy with you or with a colleague in the room.)

## Communication

OK, it's showtime. As you deliver your presentation,

1. If possible, avoid reading from a script. (Formal speeches are an exception; they generally have word-for-word scripts.) If necessary, have an outline. If possible, memorize your opening and closing so that you can maintain unbroken eye contact with audience members at those key moments.
2. Maintain eye contact. Look at one person at a time, letting your gaze linger so that they know you were speaking directly to them. Single out leaders and decision makers for extra eye contact.
3. Unless you're making a formal speech from a lectern, move about the front of your presentation area. Your graceful movements will help keep audience members engaged and attentive.
4. If your technology fails, don't try to hide the fact and don't panic or grumble. Move smoothly to your backup solution. Your preparation and grace under pressure will impress audience members. They'll appreciate how seriously you've taken this opportunity to speak with them.
5. When appropriate, close by asking for questions. After answering the last question, deliver a brief closing statement—just a few sentences, but make them as dramatic as the situation allows.
6. Consider this closing strategy: When you've finished your presentation, deliver your closing statement. Then pause, maintain eye contact and simply say, "Thank you." This strategy often triggers applause.

## Evaluation

You're done! When you're finally alone, relax and congratulate yourself. But don't get too comfortable. Before your sharpest memories of the presentation fade,

1. Ask yourself what worked well. What didn't work well? How did the technology perform?
2. Ask yourself if anything in the presentation surprised you. Was the audience responsive? If so, when? Why?

3. Analyze your clock management. Did you finish on time? Did you allow time for questions? Did you feel rushed? Or did the presentation go too quickly? If so, why?
4. Give yourself a grade. Was the presentation an *A*? *B*? *C*? Try to score a higher grade next time.
5. Learn from your successes and your failures. Build on what went well. Revise what didn't work.

As you gain experience, your presentations will become better and better. Who knows? You might even learn to enjoy them.

# Index

T - #0001 - 110722 - C378 - 254/178/20 - SB - 9780367895402 - Matt Lamination